Vanusa de Paula Santos

Socioeconomics of the Small Rural Producer:

Vanusa de Paula Santos

Socioeconomics of the Small Rural Producer:

The Case of the Santo Onofre Settlement - Pantanal de Poconé/MT - Brazil

ScienciaScripts

Imprint

Any brand names and product names mentioned in this book are subject to trademark, brand or patent protection and are trademarks or registered trademarks of their respective holders. The use of brand names, product names, common names, trade names, product descriptions etc. even without a particular marking in this work is in no way to be construed to mean that such names may be regarded as unrestricted in respect of trademark and brand protection legislation and could thus be used by anyone.

Cover image: www.ingimage.com

This book is a translation from the original published under ISBN 978-613-9-65215-0.

Publisher:
Sciencia Scripts
is a trademark of
Dodo Books Indian Ocean Ltd. and OmniScriptum S.R.L publishing group

120 High Road, East Finchley, London, N2 9ED, United Kingdom
Str. Armeneasca 28/1, office 1, Chisinau MD-2012, Republic of Moldova, Europe
Printed at: see last page
ISBN: 978-620-7-86167-5

Thank you

I thank God for everything he has given me during my postgraduate studies, for the privilege of being able to complete this research, for the development of the knowledge that has beautifully moulded itself into my being, as well as for the wisdom granted at each stage of the research.

I would like to thank my husband Gilmar Rodrigues for his support, encouragement and understanding amidst the arduous path I have travelled, my children who also cooperated by understanding their mother's absence at various times, my parents for the support they provided in the Santo Onofre Settlement, contributing greatly to the development of the fieldwork and obtaining the results, and my siblings who have always shown enthusiasm for the career I have set myself, in short, all the family members who have directly or indirectly contributed to me.

I would like to thank the residents of the Santo Onofre settlement, who gladly welcomed me and provided me with information relevant to my work, especially Aldinéia, the president of the local residents' association, who accompanied me throughout the field research.

I would like to thank the Postgraduate Department, the professors and technicians, especially my dear supervisor, Professor Dr Onélia Carmen Rossetto, who patiently taught me, guided me, encouraged me, supported me, corrected me and applauded me. I am at a loss for words to express how highly I regard her, and that she is my great role model at UFMT, I would like to thank all my classmates for the good learning moments we had together, the GECA-UFMT research group for the help and affection they gave me and for the socialisation of knowledge, affectionately contributing to the difficulties encountered along the way, I would also like to thank those who, even at a distance, supported the success of this work.

*Delight yourself also in the Lord, and he will give you
what your heart desires. Commit your way to the Lord;
trust in him, and he will do everything.*

(Psalms 37:4-5)

SUMMARY

SUMMARY

Brazil's agrarian dynamics have been inherited since the colonisation process. From the outset, land exploitation was based on the capitalist mode of production responsible for the accumulation of capital, directly influencing the growth of modernised production; however, there are those who still use the traditional system of work in the countryside, namely small rural producers. This dissertation is about the socio-economic aspects of the Santo Onofre Settlement, located in the municipality of Poconé, in the Pantanal region of Mato Grosso. The general aim of the research was to survey and analyse socio-economic issues related to the Santo Onofre settlers, as well as the main difficulties they experience, in addition to the reality of selling or abandoning plots of land. The main types of local crops were also surveyed, describing the pineapple production chain, which is considered to be the most profitable. The methodological procedures used in the research opted for a qualitative analysis of the quantitative data. The work is based on social research methodology, using semi-structured interviews with 31 local residents, resulting in data that allowed for a detailed understanding of the situational characteristics presented by the interviewees. Through the research, it was possible to ascertain the social links as well as the economic rhythm of the Santo Onofre population. Production is basically subsistence and the commercialisation of pineapples is seen as the main source of income. The need for accessory labour to supplement the family income is evident among the residents. Thus, the reality experienced in the settlement is synonymous with that of other establishments of small producers from the Brazilian agrarian reform in other regions of Mato Grosso.

Keywords: Small producer; agrarian dynamics; socio-economic.

INTRODUCTION

Brazil's agrarian dynamics have been inherited since the colonisation process. Land exploitation in Brazil was based from the outset on the capitalist mode of production responsible for the accumulation of capital. This system has a direct influence on the growth of modernised production, but this reality does not apply to everyone. Some producers still continue with the traditional system of production in the countryside, with countless difficulties, especially in Brazil.

According to Kaustsky (1986 [1899]), larger establishments arise from the concentration that is typical of capitalism. Small properties don't have the same equipment as large ones, which makes labour more arduous and requires more effort to stay on the land.

Technological advances and the tendency for large properties to expand lead small producers to experience difficulties in staying in the countryside, because if they are unable to fit into the capitalist mode of production, they tend to sell their land and leave the countryside or work in other rural establishments, or even migrate or return to the urban environment. Marx (1982) sees modernised agriculture as a necessity for the expansion of capital, thus replacing traditional practices with technically advanced methods. Therefore:

> Modern industry acts more revolutionarily in agriculture than in any other sector, by destroying the bulwark of the old society, the peasant, and replacing him with the wage labourer. The need for social transformation and class opposition in the countryside are thus equated with those in the city. The routine, irrational methods of agriculture are replaced by the conscious, technological applications of science (MARX, 1982, p. 577).

Brazil's agrarian dynamics were developed under the concentration of land, a fact that is evident in other Latin countries such as Argentina, Mexico, Paraguay and Bolivia. Based on an economic system that encourages large-scale production, latifundia became a reality from the very beginning of the formation of Brazil's territory.

The current transformations have a direct impact on the Brazilian countryside, as mechanised production methods and increased productivity are gradually replacing traditional methods. In the light of Martins' (1996) analysis, we would point out that the growth of capitalism not only redirects old relations, leaving them subject to the reproduction of capital, but also generates pre-capitalist relations, possibly the most important factor being the capitalist rent of land, since land as a natural medium, without monetary value because it is not the fruit of human labour, according to the theory has a price.

Mato Grosso, considered one of Brazil's largest producers of soya for export, is marked by the presence of latifundia. In addition to soya, the cultivation of cotton, corn and sugar cane also encourages large-scale production, further strengthening the concentration of land in the state. On the other hand, small rural producers are seeking their autonomy amidst this scenario. The struggle for land and survival on it is a notorious reality in Brazil's agrarian environment, and social movements in the countryside are important ways of demanding improvements in this area. Mato Grosso is also part of this context of the struggle for land and the search for agrarian reform, which has contributed to the development of rural settlement projects throughout the state.

The scenario of land concentration in the state of Mato Grosso today is a reflection of a past of illegal appropriation, because at the beginning of the formation of Mato Grosso, "[...] land ownership was by occupation, because there were no laws regulating the occupation of public lands, it was only in 1850 that the Land Law came into being (CARVALHO, 2012, p. 57).

Land Law No. 601 (BRASIL, Legislação..., 1850) was applied in the state for the first time, sanctioned in 1892, in order to regularise the land situation in Mato Grosso and in the same year the distribution of public lands was also supported by law, and the large Mato Grosso latifundios were thus guaranteed large portions of land, even those that did not comply with the Land Law of 1850 because they had an area greater than 3. 600 hectares achieved regularisation of ownership.(LAMERA and FIGUEREDO, 2007).

The differences in the distribution of land in Mato Grosso can be calculated using the Gini Index, which is considered suitable for inequality studies. Hoffmann (1998) mentions that the benefit of using it lies in the relationship with the association with the Lorenz curve, which serves to represent the difference between the issues analysed. The index can vary between 0 (zero) and 1 (one), where the closer it is to zero, the greater the inequality and, conversely, when it is close to one, the less differentiation between the phenomena analysed.

With regard to the Gini Index of the land ownership structure in Mato Grosso, Silva, C. (2012) mentions that, according to INCRA data, in 1992 the index was 0.813, with a decrease in 1998 to 0.763. This figure remained until 2008. Silva also cites data from the IBGE, which calculated a figure of 0.870 in 1996, while in 2006, the index fell slightly to 0.865. This shows that, despite the decline in the figures presented by both INCRA and the IBGE, the result is a slow deconcentration of land ownership in Mato Grosso, as well as the need to accelerate and implement land reform projects in Mato Grosso.

The occupation of land in the Pantanal by non-native residents stems from the process of sesmarias, which has existed since 1532, and materialised in Mato Grosso in 1727. The sesmaria system:

> [...] allowed large tracts of land to be owned free of charge by people who could prove they had the financial means to exploit them. The seasonality of the climate and the terrain, which meant that part of the properties were flooded at certain times of the year, influenced the size of the properties and legitimised the large size of the farms in the Pantanal, which often didn't have dividing fences between the establishments. Owners rarely had title deeds as a material document, but their boundaries were respected by their neighbours and marked by geographical features (ROSSETTO and GIRARDI, 2012, p. 10).

The process of using the soil for agricultural production has existed for more than 200 years, with the production of sugar cane and, as a highlight, cattle ranching, a precursor to the formation of large properties in the Pantanal, since the formation of pastures and natural salt pans provided a profitable scenario for cattle ranching in the Pantanal region, causing a large and accelerated growth of the cattle herd (MOREIRA, M. 2008).

According to Rossetto and Girardi (2010), from 1973 onwards actions were taken to regularise the land issue in the Pantanal through policies developed by INCRA, which produced a zoning of the rural environment of Mato Grosso, creating land reform projects for six regions, including the municipality of Cáceres. The authors also emphasise that in the 1980s INCRA once again directed land regularisation projects towards the Pantanal regions.

However, the situation of land accumulation in the Pantanal of Mato Grosso is still notorious, a circumstance that occurs throughout Brazil, proven by the Gini index presented by the IBGE as cited in op. cit (2010, p. 03) which "[...] indicate that in the Pantanal municipalities the Gini index went from 0.856 in 1966 to 0.872 in 2006. According to the IBGE, Mato Grosso also showed a slight deconcentration, going from 0.870 in 1996 to 0.865 in 2006."

The delimitation of the Pantanal region in Brazil, according to Vila da Silva and Abdon (1998) corresponds to the municipalities belonging to Mato Grosso: **Barão de Melgaço; Cáceres; Itiquira; Lambari d'Oeste; Nossa Senhora do** Livramento; Poconé; Santo Antonio do Leverger, while in the state of Mato Grosso do Sul they are: Aquidauana; Bodoquena; Corumba; Coxim; Ladário; Miranda; Sonora; Porto Murtinho; Rio Verde de Mato Grosso.

Of the 530 rural settlements located in Mato Grosso, 55 are in the Pantanal region, and the municipality of Poconé has 11 rural establishments that are the result of agrarian reform.

The agrarian reform settlers, whose production is based on subsistence, face obstacles inherent in the lack of structures in the areas of education, transport, financing, subsidies for production, etc. The Santo Onofre Settlement, located 15 kilometres from the Cangas District in the municipality of Poconé-MT (Fig. 1), is part of Brazil's agrarian reality, with 95 plots of approximately 15 hectares.

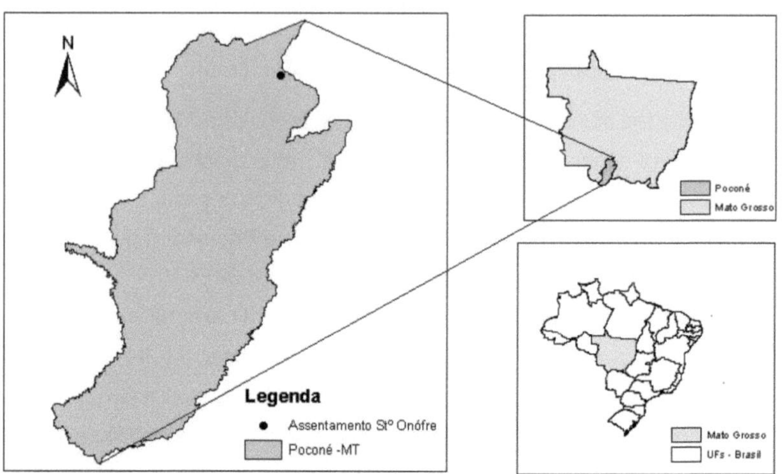

FIGURE 1: Location map of the study area.
SOURCE: SILVA, C. and SANTOS (2014).

Surrounded by large cattle ranches, the struggle to remain on the land is fierce, with those who can't stay in the countryside returning to the city, in a process that contradicts the one preached by the agrarian reform policy, because if the aim is to provide land for those who don't have it, how can we explain the renunciation of rural areas by settlers?

Of course, this phenomenon can be clarified when we analyse the way of life of peasants who are cut off from the policies that support small landowners, who, without adequate help, prefer to leave the countryside in favour of their own land.
return to the urban environment to support their families through non-rural jobs.

The art of cultivating the land in the traditional way is seen as a challenge for the settlers of Santo Onofre, because the difficulties, such as the lack of support in the technical area such as assistance and guidance for land management and the subsidies needed to produce on the plot, the commercialisation and disposal of production, the lack of adequate structure make life in rural areas arduous.

Given this situation, the aim of the research was to investigate the reality experienced by the residents of the Santo Onofre Settlement, the process of emptying and selling plots. The aim was to find out whether the settlers are helped by public policies related to the development of projects that promote income for small rural producers, access to land credit and the conditions necessary to remain in the countryside. The objectives also included questions such as: the difficulties and level of satisfaction of the settlers; and local structures. The results were obtained through field research, which resulted in research topics related to the socio-economic analysis of local residents, the main crops produced and their commercialisation, highlighting the pineapple production chain.

The first chapter presents the methodological procedures used in the research, opting for a qualitative analysis of the quantitative data. The work is based on the methodology of social research, emphasising the socio-economic aspects of the residents of the Santo Onofre Settlement. In order to better understand the results, authors who work with the methodology in question were cited.

In the second chapter, the conceptual theoretical framework was developed. The survey of bibliographies on the subject provided support for structuring the research, as well as providing a better understanding of the phenomena presented.

The third chapter deals with the agrarian question in Latin America, considering the trajectory and struggle for land in countries such as Mexico, Argentina, Paraguay and Bolivia. The processes of land concentration presented in the research took place synonymously in the countries in question, with their roots based on European colonisation, which in turn exploited the resources of the new lands, strengthening latifundia and the expropriation of smaller landowners.

The fourth chapter deals with the Brazilian agrarian question, covering issues related to the land structure, the processes of exploiting Brazilian territory, as well as the methods for obtaining and using land. The phases have been described subsequently, presenting each stage of the journey of Brazil's agrarian environment, and the results unleashed over time.

The main land reform programmes in Brazil, the projects and plans provided over time, were also surveyed, and are presented in a linear fashion, respecting the period in ascending order.

The state of Mato Grosso is part of this context, and agrarian reform projects have been developed to consolidate rural settlements throughout the territory. Problems related to land still persist in the Mato Grosso region, because the predominance of latifundia is a characteristic of the state's agrarian policy.

The fifth chapter describes the physical characteristics and agrarian reform of the Pantanal in Mato Grosso, even though it is considered a complex area, land concentration is high and small producers in this region have similar difficulties to peasants elsewhere in

Brazil. Therefore, agrarian reform in the Pantanal region is essential to give access to land to those who don't have it. This chapter also describes the study area.

This is followed by chapter six, which discusses the results obtained in the research. The main focus is on surveying and analysing the socio-economic aspects of the Santo Onofre settlement, highlighting issues inherent to the local residents' experience; the activities they carry out; their relationship with the land and their permanence on it; the types of production, describing the pineapple production chain, which is considered to be the most profitable for local residents. This is followed by the final considerations and then the bibliographical references.

The development of research related to rural areas and agrarian reform settlements has triggered a series of analyses of the authentic condition of the Brazilian peasant. The search for a decent quality of life on the land is a reality for many of those who live in the countryside.

CHAPTER 1

METHODOLOGICAL PROCEDURES.

This dissertation is based on social research, since the aim of this work was to investigate and analyse the socio-economic aspects of the residents of Santo Onofre. According to Minayo (2008), man has always been concerned with knowing the reality he experiences, using various instruments to obtain this knowledge, such as religion, art, philosophy and even science itself. Therefore, social research is appropriate in studies related to the social and economic area of man.

The work also has characteristics that are in line with the parameters of Marxist dialectics **which, according to Trivinos (1992, p.51), "dialectical materialism is the** philosophical basis of Marxism and as such attempts to find coherent, logical and rational explanations for the phenomena of nature, society and **thought"**. By drawing up a qualitative analysis, dialectical materialism provides a better understanding and appreciation of the results obtained in the field.

The construction of a research project and its development can be governed by the creativity of the researcher, because when they use new rules they contribute to the growth of science, agreeing with Minayo (2008) that there is nothing that can replace the creative capacity of the researcher.

When taking into account social and economic aspects, qualitative data is of the utmost importance and, together with quantitative data, provides a better understanding of the object being researched. Based on the studies of methodologist Richardson (1999), semi-structured interviews were carried out with 90 per cent of the residents who actually live in the area, and 10 per cent of those who live in the city and own plots in the settlement. According to Richardson (1999, p.90), qualitative research can be characterised as an attempt to gain a detailed understanding of the meanings and situational characteristics presented by the interviewees, rather than quantitative measures of characteristics or behaviours. As well as being qualitative, the research is also exploratory, and according to Richardson apud Carvalho **(2012, p. 17) "exploratory research** seeks to know the characteristics of a phenomenon in order to seek explanations of the causes and consequences of **said phenomenon"**.

To carry out the research, the following procedures were carried out: a bibliographical survey, the preparation of a semi-structured survey instrument, documentary research and fieldwork using a camera and a field notebook to survey the reality of the Santo Onofre settlement. A detailed tour of the entire area allocated to the settlement was also carried out. According to Minayo (2008), fieldwork offers the possibility of getting closer to the object of study and also of generating knowledge based on the reality present in the countryside. 31 interviews were carried out in the settlement, of which 64 were local residents, 18% caretakers and another 18% living elsewhere. The rest of the residents who were not interviewed were working or carrying out activities in the city, which interfered with the construction of comprehensive data. However, with the help of a local leader, it was possible to cover the entire area corresponding to the settlement, confirming the occupational situation of each plot.

Surveys of the bibliographies of authors who discuss the proposed topic served as support for understanding the results. According to Andrade (1997), analysing and drawing on textual references can be developed in itself or established during the construction phase of monographs, dissertations, etc.

According to Silva et al (2009, p. 04), documentary research allows:

> [...] the investigation of a given problem not in its immediate interaction, but indirectly, through the study of documents that are produced by man and therefore reveal his way of being, living and understanding a social fact. Studying documents implies doing so from the point of view of those who produced them, and this requires care and expertise on the part of the researcher so as not to jeopardise the validity of their study.

The chapters throughout the work follow a hierarchy from global to local, covering topics such as the agrarian question in Latin America, Brazil, Mato Grosso, the Pantanal and finally the presentation of the results obtained in the Santo Onofre Settlement in the municipality of Poconé, followed by the final considerations of the work. In order to better understand the results obtained in the study area, a survey of research related to rural settlements in diverse locations in the state was carried out, where it was found that peasants have similar characteristics, especially in terms of survival and the difficulty of staying in the countryside.

CHAPTER 2

CONCEPTUAL THEORETICAL FRAMEWORK

2.1- The Peasantry and the Capitalist Mode of Production.

Issues inherent to the peasantry have triggered a series of studies into their relationship with capitalism. When we analyse their economic situation, we notice that there is a certain difference in terms of land use. Kautsky (1999 [1986]) apud Girardi (2008) mentions that even though peasants are part of the market, their relations may not be capitalist. What defines the peasant is the source of their own labour, i.e. family members, while the capitalist buys labour from expropriated workers.

In order to better understand peasant relations in the face of the capitalist mode of production, it is important to understand the term *capitalism,* which according to Faria (1999) apud Moreira (2012, p. 39-40):

> Capitalism came to be the economic formation that developed the mercantile relation in the most advanced way, making it one of the constitutive relations - alongside the monetary relation (the rule of equivalence based on the fetish of abstract wealth, the law of value) and the wage-earning relation (the specifically capitalist form of exploitation of labour and production of surplus) - of this mode of production.

Capitalism is directly linked to the permanent accumulation of capital. It began in Europe, and has been noted since the lower Middle Ages (11th to 12th centuries). From the 15th century onwards, the expansion of trade due to the Great Navigations, mainly by European countries such as Portugal and Spain, further strengthened the process of capital accumulation, generating wealth through the exploitation of land, establishing trading posts around the world, this moment became known as the period of Commercial Capitalism. (LUCCI et. al., 2010).

Still according to the author, in the second phase, industry played an important role in consolidating capitalism, with the First Industrial Revolution, with the invention of the steam engine in 1769, industrial production gained great momentum. Workers were concentrated in the same place of production (the factory), consolidating the form of wage labour. At the end of the 19th century, the use of electricity and petrol as energy sources accelerated the process of production and the flow of goods, expanding the capitalist model across the globe.

With regard to issues related to peasants, the concept gained ground in the Brazilian Human Sciences from the 1950s onwards, underpinning studies on the Agrarian Question in Brazil, while in the 1970s the concept of small-scale production began to be used as an alternative to that of peasant.

Authors such as Oliveira (1991), among others, use the concept of the peasant as being based on family labour to support its members, without having a capitalist commercial link.

The concept of the peasant is analysed in the light of the Marxist M-D-M (commodity-money-merchandise) logic, where the commodity produced on the land is converted into money that is used to buy other commodities necessary for family subsistence, those that are not produced on the property.

Commercialised production aimed at profit is part of the Marxist D-M-D' logic, **which, unlike the first logic presented,** invests **money in order to** obtain merchandise, returning in the form of money and profit, thus characterising the capitalist mode of production in the countryside. According to Paulino (2010), capitalism in the countryside is based, according to Marx, on three structural pillars: land, where income is acquired; labour, which generates wages; and capital, which results in profit.

Thus, when land becomes a target for profit-making, it loses its characteristic of **"place", which** Shanim apud Paulino and Almeida (2010) consider to be a **"way of life", with economic, social and family development** as inseparable relationships.

In the last decade of the 20th century, the concept of family farmer was used as a substitute for the others. Girardi (2008), quoting Abramovay, refers to the differentiation between the concept of peasant and family farmer. In the peasantry, there is a personalisation of social ties, where the peasant is characterised as backward, poor, dependent and outdated, retrograde and peripheral in the countryside, being subject to submergence, undergoing a transformation, becoming a new category adopting another direction in production relations. On the other hand:

> Technical dynamism, the capacity for innovation and complete integration into markets would be characteristics of family farming. In developed countries, where capitalism has reached a higher level of development than in underdeveloped countries, the system itself would annihilate the peasantry and have family farmers as the main social base for development. The "metamorphosis" of peasants into family farmers would occur within this capitalism with a higher degree of development and with strong state intervention in the structuring of national markets (GIRARDI, 2008, p. 100-101).

The author also quotes Abramovay as saying that the peasant's economic rationality is essentially incomplete, because the peasantry is not an economic type, but a way of life in which social ties are personalised. The author also believes that flexibility is not synonymous with independence, but rather a reflection of the imperfect market in which the peasant is inserted, and that the peasantry is incompatible with market environments.

For Fernandes (2005), the family farmer can be considered a synonym for peasant, based on family production, a characteristic that unites peasants all over the world. Therefore, the term "peasant" and "family farmer" coexist in the literature on the agrarian aspects of Brazilian territory, however, both take on different meanings and are more related to the political stance of the authors than to significant conceptual differences. This discussion is of great importance since agrarian reform in Brazil is a recurring and contemporary theme.

Law No. 11.326 of 24 July 2006 (BRASIL, Legislação...,2006), which establishes the guidelines for the formulation of the National Policy for Family Farming and Rural Family Enterprises, does not stipulate any differentiation when referring to small landowners, so family farmers and peasants are considered under the same requirements.

According to the MDA (Ministry of Agrarian Development) website, based on Law 11.326, the requirements for being considered a family farmer are: not owning an area larger than four fiscal modules, each module being equivalent to 50 hectares, using predominantly the family's own labour for economic activities on the property, the family's income coming predominantly from activities carried out on the property, and the property being run by the family itself.

Law 11.326 (BRASIL, Legislação..., 2006) does not separate the family farmer from the peasant, and it can be considered that the scope of the law is correct, since both the peasant and the family farmer (in this case considering them as antonyms) have similar characteristics as mentioned in the previous paragraph.

There is also the thought that capitalism destroys the peasantry, but recreates it:

> [...] an intense process of disintegration of the peasantry within capitalism, but not its disappearance, because it is recreated. Capitalism has made the feudal, self-sufficient peasant disappear, but it guarantees the existence of a subordinate peasantry in a constant process of disintegration and re-creation. (KAUTSKY, 1986 [1899]).

Peasants who sell their labour or perform other types of services outside of the property carry out what we call accessory work. Paulino and Almeida (2010) refer to off-farm work as a way of increasing the family income to support the household.

For Lenin (1985 [1899]), who based his work on the development of capitalism in Russia on a period that still had strong feudal ties and was undergoing a transition to capitalism, he emphasises that the peasantry is subordinate and dependent on the capitalist mode of production, thus dividing the peasant into three groups: the rich, the middle and the poor, where the rich can become capitalists, the middle can become rich or poor, and the poor is the one who does not have enough return to maintain his family and production, having to look for other work to complete his income, this is obtained through wage labour.

Among the difficulties that smallholders can face as a result of capitalism, we can consider that impoverishment is inherent to the concentration of capital in certain poles, to the detriment of regions where disintegrations have taken place. (SHANIN, 2005), still in this argument Girardi (2008) points out that the marginalisation of peasants occurs because they do not become capitalists or workers, nor are they considered merely poor, their existence is joined to the capitalist economy in a marginalised way.

Something important to take note of with regard to the peasantry are the considerations regarding their relations, even if they are contrary to capitalism, showing that the countryside is not just a means of producing goods, but goes beyond that, it is a place of life, showing a sense of well-being, even if the peasant encounters various barriers to remaining in the countryside, According to Kautsky (1986 [1899]), the large establishments do so because of the concentration that is typical of capitalism, while the small ones don't have the same facilities as the large ones in terms of access to loans, lower interest rates, cheap transport and direct negotiation without the presence of middlemen, etc.

Chaianov (1981[1924]), commenting on the peasantry, mentions that it can be considered a non-capitalist means of production. The author also argues that the peasant unit is flexible, making greater or lesser efforts to meet the needs of the family, so what Chayanov tells us apud **Paulino and Almeida (2010) is that the number of "mouths" will govern** the effort to produce in greater or lesser quantities.

Still considering Paulino and Almeida's quote about the limits of the peasant economy, the land is not just labour land, it is also the **"dwelling place of life", a place for pets, the** orchard, the vegetable garden and the garden, where the family group reproduces itself through self-consumption.

On this subject, Amim and Vergopoulos apud Paulino and Almeida (2010) mention that the small peasant owner does not really behave like a tenant or a capitalist entrepreneur. Initially, he is obliged to produce, regardless of the market situation, at the risk of not surviving under the capitalist model of production.

In this way, peasant production is based on the circulation of goods, as Oliveira explains:

> In peasant labour, "part of peasant production goes into the direct consumption of the producer, the peasant, as a means of immediate subsistence, and the other part, the surplus, in the form of merchandise, is commercialised." (OLIVEIRA, 1986, p. 68).

Still in the analysis of rural areas, there are paradigms that seek to substantiate the relationships built in the countryside. Fernandes (2005) identifies two main paradigms: the Paradigm of the Agrarian Question (PQA) and the Paradigm of Agrarian Capitalism (PCA). The main point of discussion between the two is their position in relation to capitalism. The PQA seeks to analyse the conflicts and inequalities generated by capitalism in the countryside. The PCA seeks to understand the best ways for family farmers to integrate into the capitalist system, one of which is to increase crop productivity, reduce production costs, obtain off-season production in order to enjoy better market prices and thus have a higher income.

Each of the structures mentioned above reinforces the need to really implement a precise agrarian reform, which not only allows families to settle in the areas earmarked for agrarian reform settlements, but also provides the other necessary subsidies for production on these new properties, as well as sufficient structures for them to remain in the countryside.

Despite the claim **that "peasants" are tending to disappear,** we believe that they are still fighting to gain access to land and to remain on it in many parts of Brazil, including the state of Mato Grosso and the Pantanal region in the municipality of Poconé. In this work, we opted for the concept of peasant when analysing the small rural producers of the Santo Onofre Settlement, agreeing that the peasantry is not explicitly inserted in the capitalist logic, but in the peasant logic, noting the **distinction between the two, one being "work land" and the other "business land".**

CHAPTER 3

THE AGRICULTURAL ISSUE IN LATIN AMERICA: the case of Mexico, Argentina, Paraguay and Bolivia.

In order to address issues related to the Brazilian agrarian question, it is important to have knowledge of the subject in other countries around the world. In this respect, the relationships that exist in rural areas in Latin American countries provide relevant information for understanding Brazil's agrarian scenario.

The agrarian question in Latin America reflects a synonymous reality in different Latin American countries. The struggle for land has triggered countless processes over the years, starting with colonisation in countries such as Mexico, Argentina, Paraguay, Bolivia and Brazil. The peasant lives the reality imposed by the capitalist mode of production, and in order to be included in this economic model, the peasant faces several obstacles.

The search for the right to land is essentially the same in the most diverse countries of Latin America: the desire to cultivate the land and to have **a "place" to live makes them similar:**

> Every man, in fact, has the right to aspire to the ownership, for himself and his family, of a piece of land, on which he permanently establishes his home and from which he derives, through labour, his own sustenance and that of his family. He has the right to aspire to ownership and not merely to temporary use of the land [...]. For land is the necessary fulcrum of all human activity and the ultimate stage of every economic endeavour (VICENTE SCHERER apud STEDILE, 2005, p. 62).

At the outset, when dealing with issues related to the agrarian question, it is worth pointing out that there are various ways of approaching this issue. Numerous studies can be carried out, because depending on the area of study, the view of the issue varies. Sociology, according to Stedile (2005), uses the concept of the agrarian question to explain the ways in which social relations are developed in the arrangement of agricultural production.

In Geography, the concept of the "agrarian question" is most often **used in an** attempt to explain the way in which human beings appropriate and use land, as well as human occupation of the territory. In history, it is used to help explain the progress of the class struggle for command and control of territories and land ownership.

Some Latin American countries have carried out agrarian reform projects that have worked positively, in an attempt to satisfy the desires of the peasant population. It is clear that some of these countries are in the adjustment phase, experiencing the reality of expropriation, especially during the 20th century, where the rural structure of large estates, inherited from the colonial era, prevails in most of them.

Mexico has become a relative model for other Latin countries, with a dictatorial policy in which Porfirio Díaz (1876 - 1911) favoured foreign companies, forming veritable oligarchies in Mexican territory, leaving the poor and natives at a disadvantage.

The latifundios expanded their borders at this time, appropriating indigenous lands and small rural properties with the government's approval, causing an increase in the rural proletariat and generating a revolt among the population that led to the so-called Mexican Revolution in November 1910. At this time, even community and church lands were confiscated and later commercialised, further increasing the discontent of the poor, small producers and natives (FERNANDES, 2008).

Unhappy with the agrarian reality at the time, groups of peasants in the province of Morelos decided to resist the legacy of becoming wage labourers, wanting to return to **their "ejidos", or** communal lands, where everyone could enjoy the benefits collectively, thus rescuing the culture they had inherited from the ancients (DEVEZA, 2011).

One of the big names in the Mexican Revolution was Emiliano Zapata, who defended the idea of distributing land to the poor with the **slogan "Land and Freedom", which led to Diaz's departure from government in 1910 and** Francisco Madero taking his place in 1911.

In the 1920s there were forced invasions of properties by peasants considered to be rebels, (re)organising the ejidos through the Zapatista Laws of 26 October 1915 and 03 February and 05 July 1917. The government therefore called the lands distributed ejidos. It's worth noting that during this period, the best land was being distributed to the peasants, with the agricultural credit system available to them.

The Mexican Revolution brought about important democratic actions, but, as Schmidt (1996) reports, years later, agrarian reform in Mexico stopped growing and stopped in its tracks, given the ideology preached by the government that improvements would come, step by step and in order. For him, some peasants still suffer from the decrease in investment in land distribution, making it clear that the struggle for land continues, suffering the most diverse difficulties as in other Latin American countries.

In the case of Argentina, Ribeiro **(2006, p. 31) mentions that** "[...] the issue of access to land had a more perceptible influence on **government** discourses than **rural labour legislation." In the government of Juan Domingo Perón (1946 to 1955), the discourse preached in Argentina was "to produce", urging** workers to increase production. Perón put aside the old anti-landowning discourse and focused on improving and increasing production in the country. Agrarian reform in Argentina took on new relevance:

> As for the Argentinian case, our point of reference is not Perón's inauguration in 1946, but the arrival in power via a military coup of elements of the Grupo de

In Perón's government there was some concern about the rural exodus, which was treated as a danger to be corrected, because the countryside was considered a national strength. In a **speech to Congress in 1948,** Perón **"[...] affirmed that the** effort to fix man to the soil, seeking to bring wages and adequate working conditions, was also aimed at avoiding unemployment [...] **caused by the swelling of the cities." (RIBEIRO,** V., 2006).

Wahren (2008) points out that the 1990s were marked in the history of Argentina, because in this period numerous problems related to society were unleashed, in addition to the food crisis, unemployment caused by the privatisation of state-owned companies reached 18% in 1995. The alleged reason for the privatisation process was the inefficient functioning of state-owned companies, leaving countless civil servants unemployed. An example of this was the privatisation of the oil company "Yacimientos Petrolíferos Fiscales **(YPF)".**

At this time of crisis, the UTD (Union of Unemployed Workers) was formed in 1996, the year that triggered the uprisings in Cutral-Có and Plaza Huincul, General Mosconi and Tartagal (Salta). The revolts quickly became one of the most unique social protests in Argentina in the 1990s. The actors taking part in the protests were: producers and peasants; rural workers and landless tenants; teachers; civil servants and the unemployed.

Still for Wahren, in the face of the territorial and social disfigurement in Argentina as a result of the privatisation of YPF in the 1990s, UTD aims to recover the social fabric broken by the privatisations, developing self-help projects that involve community work for the region, productive projects and cultural projects.

The development of projects, especially those linked to agricultural activities, marks and reconfigures a territory abandoned by the state company YPF since its privatisation, which led to a dispute over its land and natural resources by a range of oil companies and soybean growers who wanted to take advantage of the situation. In the province of Salta, in the last decade, according to Wahren (2008), large-scale production for the export of *commodities has led to* the advance of the agricultural frontier to the north-west, growing mainly GM soya.

Food production projects emerged as a milestone in the midst of the food crisis in the 1990s in Argentina, as food production not only supplied the families participating in the projects, but also schools, nurseries, community centres, among others, partly solving a serious social problem for the government. The community gardens mainly produced vegetables, and the surplus was allowed to be sold, i.e. what was not consumed by the families or distributed in the community.

Some projects have been unsuccessful, such as the **"Cecafa" project,** which is a 350-hectare development. Acquired from the municipality with the aim of installing around thirty small producers, with plots of between 10 and 12 hectares, it has been paralysed due to recent flooding problems. Another example is the Ecological Waste Bin Project, which worked in two stages, the first collecting plastic waste from the town of Mosconi, and the second compacting the plastic and storing it in warehouses for sale. The compacting machine is semi-manual. The project is currently paralysed due to a lack of buyers for recycled plastics, and transport to other locations is not viable due to the high cost of freight.

However, according to Wahren (2008), UTD has developed numerous projects such as: seed sorting, where the company Semilla del Sur outsources part of the sorting; the portable sawmills, which use wood from deforestation, processing and recovering what would have been burnt to plant GM soya; the metal workshop, training young people for the labour market, the carpentry; the textile workshop, the brick factory and the housing cooperatives, much of the machinery and subsidies were acquired from the Ministry of Social Development of the Nation, so the positive results of many projects come from state policies and programmes, which make the effort to improve both the countryside and the cities worthwhile.

UTD, with its collective actions and self-help projects, intervenes in the political, social, economic and cultural construction of the territory, facing conflicts mainly with the oil multinationals and agribusiness that are also disputing the Argentine territory.

In Paraguay, social relations of production are considered backward in the face of the reality imposed by the capitalist mode of production, compared to other countries:

> [...] Capitalist agriculture had already been consolidated by the end of the 19th century. In Paraguay, the first clear sign of the precariousness of this form of agricultural exploitation began with the European and Japanese colonisation that settled in Itapúa from the 1920s and 1930s (VILADESAU, 2008, p. 17).

During this period, with a low population and an abundance of land, peasants organised themselves into small cooperatives to market their produce, mainly with the aim of regional marketing. From the 1970s onwards, agro-industrial agriculture grew stronger with the production of soya and wheat (VILADESAU, 2008, p. 18).

In 1960, during the government of General Stroessner, **the "Agrarian Statute" was created, also known as the "Agrarian Reform", which was the basis for the entire** colonisation **process** until the 20th century (FIGUEREDO, 2005). From then on, the "Instituto de *Bienestar Rural* **(I.B.R.)"** emerged, **the body responsible for organising the distribution of** land to all compatriots. During this period, vast tracts of land were distributed for colonisation through the Rural Roads Housing Programme, with the aim of providing good infrastructure on the roads. Figueredo (2005) also emphasises that:

> The incorporation of foreigners into agrarian activities was very noticeable at this time. The country's eastern frontiers were generally occupied by large estates, whose primary activity was the exploitation of timber for the Paraná market. This led to extensive deforestation using large machinery and chainsaws, producing large areas of land ready for agricultural and/or livestock activities.

Paraguay, with its small estates and old settlements, gradually experienced the arrival of agribusiness:

> The first significant (and devastating) expansion of capitalist agriculture, already in the hands of multinational companies linked to agro-export, took place with the expansion of the soya frontier in the southern states of Brazil in the late 1960s and throughout the 1970s (VILADESAU, 2008, p. 18).

At the same time as soya, cotton production increased considerably, making it the main crop of the minifundistas and also benefiting intermediaries and agro-exporters. According to Viladesau (2008), soya cultivation expanded in the 1980s and early 1990s and in the new production cycle between 1999 and 2000, transgenic soya increased agrarian capitalism in Paraguay.

With regard to peasants, the situation is the same as in other regions of the world, since the expansion of the agricultural frontier on peasant land has led to an increase in the number of small properties sold to agribusiness, leaving a mass of idle peasants without land to cultivate who end up selling their labour or moving to the outskirts of cities.

At the end of the 20th century, popular demonstrations broke out in Paraguay demanding social, political and economic improvements, including the land situation. At this time, organisations of small **landowners called "S/h t/e/ras"** emerged, seeking the settlement of families on their own land (FIGUEREDO, 2005).

Another problem, according to Figueredo (2005) **was that** "the low prices of cotton on the international market warned of the decline of its production, causing more and more small and medium-sized producers to abandon the activity".

As Figueredo (2005, p. 17) mentions, **"Paraguay has experienced** many socio-economic transformations throughout its history. With a large-scale economy based on the primary sector, more specifically **agriculture."** Soya cultivation is increasingly gaining ground in Paraguayan territory, leading to the impoverishment of many peasants who are not part of this system. An initiative on the part of some small and

medium landowners is the formation of co-operatives and loans, improving productivity and technifying the mode of production so that they can insert themselves into the capitalist mode of production.

Like other Latin American countries, Bolivia has also been undergoing considerable changes in its territory. From the beginning of its colonisation by the Spanish, Bolivia's main economic exploitation was mineral resources, and when mining sources were exhausted, oil and natural gas became the substitute.

A landmark moment in Bolivian history was the agrarian restructuring of 1952, considered the second major social revolution in Latin America. The colonisation policy from 1950 onwards caused profound changes, and Bolivia also took a real settlement initiative towards the east in 1953, a time when the country underwent intense political, economic and social changes (PINTO and FIGUEREDO, 2011).

According to Lopes apud Pinto and Figueredo (2011), the investments made by the government were precarious, with no infrastructure or conditions, and the peasants were sent to a totally unknown region, which made it difficult to structure these families in the new rural establishments. The land reform process in 1953 brought about important changes in the Bolivian countryside, Albó apud Pinto and Figueredo (2011, p. 24) mentions that:

> The introduction of the rural union on peasant and indigenous lands would mark the subsequent organisation of the ayllus and markas, whether through its congruences or discrepancies with their traditional structure. At the same time as providing a cohesive body for the many communities, the 1953 agrarian reform, often by expropriating the haciendas from colonisation areas, worked towards returning the land to peasant families in the form of small individual family farms [...]. These transformations in the status of the land interfered in everyday life, stimulating (and stimulating) various conflicts among the communities over property boundaries and even a new concentration of land, based on the purchase of individual properties from small farmers who had no way of cultivating their own plots, thus causing a certain internal division within the traditional communities.

The foreign colonies that settled in Bolivian territory date back to the 1950s, working mainly in intensive agriculture.

of great importance for the production of wheat, soya and dairy products. Most of the investments made by these settlers came from the countries of their origin, and major changes occurred in the Bolivian landscape due to the practice of agriculture.

The revolution that took place in the country in 1952, the development model directly affected the economy, politics and social relations, giving rise to:

> [...] the application of fundamental measures that attacked the distribution of wealth and aimed to incorporate workers and peasants into institutional life. The measures implemented were: the nationalisation of mines, land reform and universal suffrage (TEODOVICH apud PINTO e FIGUEREDO, 2011 p.24).

Idem, Latin America can be considered the region that has achieved the most far-reaching reforms in its trade policies. Agribusiness, especially soya cultivation, has had an impact on the economies of several Latin American countries, including Bolivia, which according to Pinto and Figueredo (2011) is the fourth largest soya producer in South America.

In this context, family farming was directly affected, as the incentive to increase soya production following the climatic problems that hit the United States meant that large estates tended to expand production for export.

After the 1990s, soya cultivation increased, with a rise in the number of small soya producers who, due to the crisis in traditional agriculture, surrendered to cultivation for capitalist agriculture. As a result of the increase in imported food, Bolivian peasant agriculture had its economy shaken, unable to compete with the low **prices of imported products, and** "as imports grew, prices fell. Between 1985 and 1989 there was an average drop of 30 per cent in the prices of peasant agricultural products" **(op. cit.**, 2011, p.30).

In the 1970s, social movements began to exert a direct and indirect influence. The National Revolutionary Movement (MNR), with its presidential candidate Paz Estenssoro, presented a pro-revolutionary discourse, preaching a possible restoration of Bolivia, and came to power in 1985,

where, according to Lemgruber (2006), he took a firm stance on the unions, confronting them and pursuing a harsh economic policy that led to unemployment, especially among miners.

Representing the left, the COB (Bolivian Workers' Central) was weakening more and more, carrying out population mobilisations to stop the mines from closing, but it didn't achieve any results. Since 1965, the COB had already been working on social movements, decreeing a general strike against restrictions on trade union freedoms, which led to the army intervening in the mines with the arrests and banishment of leaders. Thousands of workers were sacked and wages frozen despite the rising cost of living. Small producers in Bolivia suffered directly, many of whom, unable to bear the high cost of living in the countryside, joined the ranks of the idle in Bolivian territory.

Another important social movement in Bolivia from the 1990s onwards was the **"cocalero"** movement:

> [...] refers to the coca-growers, organised around the Federations of Coca Leaf Producers of Yungas and Chapare. The organisation of these actors is related to the situation experienced by the miners during Bánzer's dictatorial regime. Expelled from the mines, the miners migrated to the coca plantations in the Bolivian valleys, in the Yungas and Chapare regions, and dedicated themselves to growing the leaf. (LEMGRUBER, 2006, p. 04)

At this point, the figure of Evo Morales appears as a great indigenous leader of the coca-growers' movements, applauded by the people and supported by the media,

protesting against the situation experienced by the Bolivian people, gaining momentum for future electoral campaigns in his country, Morales, supported by MAS (Movimiento Al Socialismo), became president of the republic in 2006.

On taking office as president of Bolivia, Morales developed the "Land Plan", which is an agrarian reform policy:

> [...] it provides for the immediate distribution of 2.2 million hectares of land not granted by the government, the so-called "tax lands", in favour of peasants, indigenous people and native peoples who do not own or own an insufficient area of land, especially in the northern Amazon. (LEMGRUBER, 2006, p. 06).

The Bolivian regions that have received the most attention from land reform policies in the northern Amazon are the Departments (States) of Pando and Beni, bordering Brazil with the States of Rondônia, Amazonas and Acre. In addition to distributing land, the president also combated latifundia, especially in the Department of Santa Cruz, where land concentration was prevalent, by establishing rules regarding the number of hectares each property could own. This is how the land is distributed:

> [...] not because some are unproductive, as in Brazil, but because they are over 5,000 hectares. The new Bolivian Constitution, endorsed by the population with more than 60 per cent of the vote, bans larger landholdings. [...] One of the things that the Bolivian Constitution stipulated was the ban on large estates (GOMES, 2009).

Morales has also defended sovereignty over natural resources since the beginning of his political career, given that Bolivia has one of the largest reserves of natural gas in South America and that Brazil is the largest consumer market for the product.

According to information obtained from the MRE (Ministry of Foreign Affairs) website, the pipeline was officially inaugurated on 3 February 1999, in the presence of the president of Bolivia and Brazil, Hugo Banzer Suárez and Fernando Henrique Cardoso respectively. According to the MRE, the main aim of the gas pipeline project was to increase the share of natural gas in Brazil's energy matrix. The pipeline runs from Santa Cruz de La Sierra (Bolivia) to Campinas (SP-Brazil), covering a length of 3,417 kilometres.

However, in 2006, there were major changes in relation to the Bolivia-Brazil gas pipeline, with the promulgation of the hydrocarbon nationalisation decree on the first of May, in order to fulfil the promises made during the election period regarding sovereignty over Bolivia's natural resources (BARUF et. al., 2006).

Despite public policies that favour peasants and small landowners to a certain extent, Bolivia still faces serious problems related to this class of society, not to mention the large number of people living in precarious situations. Bolivia is considered the poorest country in South America, with a high percentage of its population living below the poverty line. This scenario requires even more help from the Bolivian government and the launch of more land reform policies.

CHAPTER 4

THE BRAZILIAN AGRARIAN QUESTION.

The Brazilian agrarian question has been the subject of debate and countless studies. Analyses related to the field are extremely important for understanding the phases of agrarian reform in Brazil. We consider the concept of the agrarian question to be: "[...] the set of interpretations and analyses of the agrarian reality, which seeks to explain how the possession, ownership, use and utilisation of land is organised in Brazilian society." (STEDILE, 2005, p. 15).

We consider the Brazilian agrarian question from the perspective of the Paradigm of the Agrarian Question (PQA), understanding that:

> The PQA analyses the countryside on the basis of Marxist theory and the central axis of discussion is land rent, the process of differentiation and recreation of the peasantry, conflict and the negative consequences for the peasantry of the development of capitalism in the countryside. For the PQA, the development of peasant agriculture depends on solving these *problems,* which requires going against the general laws of capitalism (GIRARDI, 2008, p. 92).

Brazil's agrarian formation is based on the exploitation of the land, knowing that:

> From the dawn of our society until 1500 AD, history records that the people who inhabited our territory lived in social groupings, families, tribes, clans, most of them nomadic, basically dedicated to hunting, fishing and fruit extraction, partially mastering agriculture. (STEDILE, 2005, vol. 01 p. 18)

With the arrival of the Portuguese in 1500 in our territory, the dynamics experienced by the natives changed completely. The search for products of **commercial value triggered a new rhythm for the "new land" discovered.**

Thus, in this process of invasion, they achieved domination of the entire territory, causing the natives who already lived here to join their production regime, albeit in a forced way.

The capitalist mode of production at this time was based on trade, in other words, characterised by the mercantile capitalism that predominated in Europe, everything revolved around making a profit and accumulating capital through trade. Of all the production acquired on Brazilian soil, large parts were exported, making the Portuguese crown ever richer. Thus:

> The organisation of Brazil's agrarian space began with colonisation. Extensive areas of land were donated by the king of Portugal for the cultivation of sugar cane in exchange for the payment of a sixth of the production obtained. This was the Sesmarias Law, which was also implemented in other Portuguese colonies. The distribution of land in Brazil therefore began with the monoculture latifundia, geared towards the foreign market (LUCCI et al, 2010, p. 326).

Still on this issue, Rocha et al (2010, p. 01) contributes by quoting that:

> Brazil's territorial history began in Portugal with the sesmarias Law of Dom Fernando I (13th century). The word sesmaria comes from sexmo, and is equivalent to the territorial area divided and drawn by lot to each municipality for cultivation and enjoyment, for a certain period of time. [...] Sesmeiros were the municipal magistrates in charge of the distribution of land in the alfoz.

The sesmaria in Portugal was an attempt to boost agriculture, which was going through times of crisis, thus avoiding the emptying of the countryside and the increase in the urban population. The newly discovered land (Brazil) was important to exploit in order to make up for the losses in Portugal's agricultural sector.

The sesmaria system began on Brazilian soil in 1530, with the arrival of Martin Afonso de Souza with an authorisation **that "allowed him to** grant sesmarias of the lands he found that could be used." **(ROCHA** et al 2010, p. 01). According to the situation presented, Brazil was influenced by the accumulation of land from the very beginning, ever since the first governor general was appointed:

> [...] its regiment of 17 September 1548 brought about the transformation that would slowly take place in the sesmarias legislation, under the influence of the colonial environment. The landowning spirit came into force, with men of possession (future plantation owners and

(ROCHA et al, 2010, p. 01-02).

Another situation in the early Brazilian period was the creation of the Hereditary Captaincies, which at first had the main objective of preventing the loss of land to France, because the Portuguese spotted French ships loaded with brazilwood, which signalled a French invasion of our territory. Thus:

> In March 1534, the King of Portugal, Dom João III, divided the country's coast into Hereditary Captaincies. There were fifteen plots that made up twelve captaincies, stretching from the island of Marajó in the north to the south of the state of Santa Catarina. They were defined as linear strips of land, which ignored geographical features, and ran from the coast of Brazil to the Treaty of Tordesillas. Therefore, initially only 20% of South America belonged to Portugal under this treaty, which determined as Spanish the lands situated beyond 370 leagues west of the Cape Verde Islands (one league being the equivalent of 5.9 kilometres). (INNOCENTINI, 2009, p. 16).

The limits established by the Treaty of Tordesillas were not respected and Portugal increasingly advanced into Brazil. The captaincies had the duty of populating the interior of Brazil, guaranteeing possession of the land for Portuguese rule. This feat, that of populating the Brazilian interior, was intensified with the discovery of important minerals such as gold. With mineral extraction, especially gold, towns and villages were quickly formed, as in the case of Cuiabá/MT. Below is table 01 on the first Hereditary Captaincies in Brazil with their respective captains:

Chart 01: Captains of the Hereditary Captaincies: the first division of donated plots in Brazil.

Captains	Location
João de Barros and Aires da Cunha, First Quinhão	Maranhão
Fernão Alvares de Andrade	Maranhão
António Cardoso de Barros	Ceará
João de Barros and Aires da Cunha, Second Quinhão	Rio Grande
Pero Lopes de Souza, Third Quintion	Itamaracá
Duarte Coelho	Pernambuco
Francisco Pereira Coutinho	Bahia
Jorge Figueiredo Corrêa	Ilhéus
Pero do Campo Tourinho	Porto Seguro
Vasco Fernandes Coutinho	Holy Spirit
Pero de Goés	St Thomas
Martim Afonso de Sousa, Second Quinhão	Rio de Janeiro
Pero Lopes de Souza, First Quinhão	Santo Amaro
Martim Afonso de Sousa, First Quinhão	St Vincent
Pero Lopes de Souza, Second Quinhão	Santana

SOURCE: Innocentini (2009), Organised by SANTOS (2013).

The system of captaincies was divided up and renamed several times, the last one being the Captaincy of Sergipe in 1820. The system was formally ended on 28 February 1821, a year before Brazil's declaration of independence, when most of the captaincies became provinces (INNOCENTINI, 2009).

When it came to land appropriation in Brazil, the concentration of land remained in the hands of the landowning elite, so the system of Hereditary Captaincies contributed to this unequal process of land ownership in Brazil, which has had consequences right up to the present day.

With regard to the privatisation of land in Brazil, it began after 1850, when English pressure for the abolition of slavery in Brazilian territory led to the first land law, which gave land a price (LUCCI et al, 2010). From then on, property became private, with the right for owners to buy and sell.

The former slaves, unable to acquire any land, became wage labourers. A large contingent moved to the cities because they didn't want to submit to the same bosses, while others remained in the countryside working for the big landowners.

According to Stedile (2005, p. 23), Law No. 601 of 1850 "[...] **was the baptism of latifundia in Brazil"**. It regulated and consolidated the model of large rural property, which is the legal basis, to this day, for the unjust structure of land ownership in Brazil.

Legally, slavery in Brazil ended with the enactment of the Golden Law in 1888, but many former slaves left the countryside for the cities, forming peripheries that favoured the emergence of favelas in Brazilian cities.

The Land Law therefore marked a period of transition in Brazil, leaving behind the slave system and moving towards the privatisation of land, which now had market value and could be freely sold:

> [...] the Land Law of 1850, as is well known, can be considered the landmark of the transition from colonial society's form of land appropriation to modern land ownership. It was created in a context in which the slaves were expected to be freed in the near future, which could put private property at risk, since until then there had been no legislation limiting land ownership, only the imperial power regulated access to the right to farm the land, thus playing a fundamental role in the process of transition from slave labour to free labour. (MOREIRA, M., 2012, p. 35-36).

The 19th and early 20th centuries were also marked by the arrival of millions of free European immigrants to Brazil, promised fertile and cheap land by the Portuguese crown. The immigrants who settled in southern Brazil received between 25 and 50 hectares of land. A large contingent of immigrants went to work on the coffee plantations in the state of São Paulo, thus characterising a migratory process of different forms across Brazilian territory.

Brazil went through the so-called colonato regime, named in this way by sociologists:

> [...] was the establishment of specific social relations in coffee production, between farmers and settlers, [...]. Under this system, the settlers received the coffee plantation ready, previously formed by slave labour, were given a house to live in and the right to use an area of approximately two hectares per family to grow subsistence products and raise small animals, thus achieving better conditions for survival. (STEDILE, 2005, p. 26).

Martins (1990), when dealing with issues related to the expansion of coffee plantations in the interior of São Paulo, points out that this phenomenon occurred through the exploitation of the labour of mainly European immigrants, who were available from the opening of the land with the clearing of the native forest, to the clearing of the land.

planting coffee plantations, receiving in return permission to cultivate a certain portion of land for farming and raising diversified animals to feed the family.

According to Martins (1990, p. 62), the settler was seen as capitalised income for the landowner, **because "[...] he had** to pay for transport, food and the settling in of the settler and his family [...]... labour continued to take the form of capitalised income for the landowner."

The work carried out during the settlement period differed from that carried out during the slavery period: the settlers used the labour of family members, i.e. the husband, wife and children over the age of seven, while the slaves worked in groups, thus socialising the tasks of the settlers' families:

> [...] he preserved the "individuality" of his work. He received a parcel of the coffee plantation with the task of keeping it free of weeds [...]. He was also in charge of harvesting the coffee and this was where family labour became more intense [...].

> Each family was given a certain number of coffee trees to take care of, based on
> 2,000 trees per adult male worker. Women and minors over the age of 12 could
> handle 1,000 coffee trees. [...] the settler's annual monetary income depended on
> the degree of labour intensification he could impose on his family.
> (MARTINS, 1990, p. 82).

From 1930 onwards, there was a crisis in the agro-export model, and with the development of industry in Brazil during this period, the agrarian issue was characterised by the growth of industry, reflecting a certain economic dependence on the part of the agricultural sector.

The advance of the industrial sector resulted in the so-called rural exodus, with a large mass of peasants leaving the countryside to work in industries in the cities, thus many expropriated peasants, deluded by the new form of labour, left the countryside to work in industry (LUCCI et al, 2010).

With high industrial demand and the emergence of a new form of income, large masses of rural labourers settled on the edges of cities, in precarious infrastructural conditions - the so-called urban peripheries.

In addition to the arrival of immigrants in Brazil, another process contributed to the emergence of the Brazilian peasantry, based on the mixed-race populations made up of whites, blacks and Indians, formed over 400 years of colonisation (STEDILE, 2005).

During the same period, the Northeast Peasant League emerged in Brazil as a movement of peasants who united in an attempt to fight social exclusion and defend the right to access land, combating the concentration of land in the Brazilian Northeast. The movement intensified after the 1940s and "was permeated by conflicts between peasants and landowners". (SILVA, T., 2009, p. 01).

The Northeast Peasant League, along with other rural movements throughout Brazil, were important in pressurising the government to demand changes in the Brazilian agrarian structure at that time, and according to Oliveira apud Silva, T. (2009, p. 11) the peasant movement in the Northeast was of a national nature " [...] from a state of tension and injustice to which peasants and salaried workers in the countryside were subjected and the profound inequalities [...] .

From the 1930s until **1945, Brazil was known as the "Vargas Era", when Getúlio Vargas ruled the country. With all the** changes in Brazil's history, land ownership remained concentrated in the hands of a few and landowning power continued to reign.

During the Vargas era, the increase in the industrial sector in Brazil, encouraged by the capitalist mode of production, led to the emergence of an industrial sector linked to **agriculture, the so-called agro-industry "which was the implementation of the**

industry for processing agricultural products." (STEDILE, 2005, vol. 01, p. 29).

With the new model in force in Brazil, the class of the agrarian bourgeoisie and large estates grew even more, leaving small producers increasingly excluded from the capitalist agrarian scene. What was left for the peasants was to provide cheap labour for the urban industries, produce low-cost food for the cities and produce agricultural raw materials for the industries.

In 1963, the Rural Workers' Statute was instituted, with the sanctioning of Law 4.214 of 2nd March 1963 (BRASIL, Legislação..., 1963), and the Rural Workers' Assistance and Welfare Fund was also created, undergoing adjustments in 1967 and 1969 which restricted:

[....] the social security action plan, [...] until the Complementary Law of 25-05-71 instituted the Rural Worker Assistance Programme - Pro-Rural, whose regulations were approved on 11-01-72. Pro-Rural provides for old-age and disability pensions, pensions and funeral aid, as well as health and social services [...] (FERRANTE, 1975, P. 191).

Still according to Ferrante (1975), the Statute was established in the midst of a scenario of conflicts related to the problems of the Brazilian countryside, but it cannot be analysed as a concrete attempt to solve rural problems, but rather as a mechanised ideology in an experiment to contain the revolts of the peasant masses.

The development of the Statute was conceived with a certain lack of interest, even from those on the political left, in a way catering to the industrial bourgeoisie, which was waiting for labour laws to be implemented in the countryside in order to turn rural workers into consumers of industrial products.

Another important moment occurred in 1964, when the period known as the Military Dictatorship actually began, when the Land Statute was created. At this time, the agrarian reform implemented did not have beneficial results for the rural environment. The effects were harmful for two reasons: the lack of implementation of the agrarian reform, which did not get off the drawing board, and the way it was established in the **Rural** Workers' Statute, **which** "encouraged the expulsion of workers who still lived under different forms of production relations" **(MANGOLIN, 2011, p. 17), as well as promoting the departure of a large** mass of workers from the countryside towards urban areas.

According to the MST (2010), the Brazilian military regime was considered strict and violent towards the peasants, and together with the whole nation they were deprived of the right to express themselves in the form of demonstrations and to hold meetings to organise protests in favour of land reform. On this occasion:

[...] the dictatorship implemented a more concentrated and exclusionary agrarian model, installing a selective agricultural modernisation that excluded small-scale farming, boosting the rural exodus, the export of production, the intensive use of poisons and concentrating not only the land but also the financial subsidies for agriculture (MST, 2010, p. 09).

In the years to come, especially in the 1970s, the federal government **implemented public policies to encourage the occupation of "empty spaces" in the** Brazilian interior, with the aim of easing agrarian conflicts in the south and northeast of **Brazil. With the "March to the West" programme, there was a** considerable **population increase in the** central regions of Brazil, which resulted in the emergence of several municipalities in the state of Mato Grosso, such as Nova Mutum, Lucas do Rio Verde, Sorriso, Sinop, among others. According to Carvalho (2012), the "March to the West" took place through colonisation projects and the establishment of the city of Brasilia, among other government security programmes.

Rural social movements in Brazil intensified from 1985 onwards, most notably the Landless Rural Workers' Movement (MST), founded in 1984 at a meeting of peasants and landless people in Cascavel in the state of Paraná (MST, 2010), demanding fair agrarian reform, assigning land ownership to those who don't have it, as well as public policies to encourage people to stay in the countryside. Conflicts have broken out in the **countryside between the "landless" and landowners, and the struggle for land has caused,** and still causes, deaths and other violent situations, reflecting a past of large property formations. The MST emerged in 1984 at a national meeting of rural workers in Cascavel in the state of Paraná with the **objectives of "fighting for** land, fighting for agrarian reform and fighting for **social** change in **the country" (MST, 2010, p. 09).**

On 10 October 1985, the I PNRA (National Agrarian Reform Plan) was created with Decree No. 91.766 (BRASIL, Legislação..., 1985), a plan presented by MIRAD (Ministry of Agrarian Reform and Development) in the government of José Sarney, and the execution of the plan was carried out by INCRA, in accordance with the proposal presented in the Decree:

> Now is the time for action. And this need to fulfil the nation's aspirations is not just a result of the constitutional imperative, the formal commitment of the Democratic Alliance and the choice made by the European Parliament.
>
> Government for firm action in the social field. As President José Sarney said, it was a question of redeeming a social debt owed to millions of rural workers and also of offering a response to Brazil's challenge to its own destiny. (BRASIL, Legislação..., 1985)

Other programmes and proposals for land reform in Brazil have been **presented, such as the "Agrarian Programme" developed by the** Workers' **Party** (PT) in **1989, with the slogan "Nothing will be as it was before", attributing** government measures for land reform that were considered essential for the constitution of a democracy that would benefit society. However, only 82,690 families were settled, whereas the plan was to provide land for 1.4 million families (MST, 2010).

With the start of President Fernando Collor de Mello's administration in 1990, there were no developments in rural policies. It was only during his vice-president Itamar Franco's administration in 1993 that the Agrarian Law (Law 8.629) was regulated, but it did not consolidate expropriations for agrarian reform. From 1994 to 1998, under the government of Fernando Henrique Cardoso, the rural exodus increased, because a large number of small farmers were in debt to the banks. Also during this period, the Banco da Terra (Land Bank) was set up, which financed the purchase of rural properties to set up settlements, and the payment of the plot to the settlers was made in instalments.

In the government of Luiz Inácio Lula da Silva, the PNRA was reformulated and presented as the II PNRA. According to the MDA's website, the aim was to guarantee land to 530,000 families by 2006, with 400,000 families settled through the agrarian reform programme and another 130,000 through the National Land Credit Programme. According to the MDA website:

> The new PNRA works with the concept of territorial development. The aim is to put an end to the idea of a single settlement model to be adopted throughout the country, but rather to establish and develop settlements according to the potential and characteristics of each region.

Among the Lula government's policies for sustainable rural development and solidarity is the PNRA, which, as Stedile (2005) cites, was the start of a concrete agrarian reform programme in Brazil.

In the current period, according to Brasil apud Moreira (2012), the country has 4,367,902 family farming establishments, totalling 84.4% of the number of rural establishments, but in terms of the overall area of Brazilian rural establishments, this corresponds to just 24.3%. In contrast, large properties account for 15.6 per cent, occupying 75.7 per cent of the total land area. According to Barcellos, agrarian reform in Brazil is currently carried out through:

> [...] the purchase or expropriation of private estates considered unproductive by the Federal Government, in various areas of the federation, and by the National Institute for Colonisation and Agrarian Reform (INCRA), which distributes and allots these lands to the families that receive them. (2013, p. 01)

Barcellos (2013) also mentions that, according to IBGE data, in the last 20 years there has been no considerable growth in properties of less than 10,000 hectares, and according to the Agricultural Census, these properties account for a total of 2.7% of all agricultural land, while 46% of land is owned by 1%, thus confirming the reality of land concentration in Brazil. Data from INCRA shows a total of 659,184 agrarian reform settlers, a figure that is considered insufficient in the face of the large population waiting for an agrarian reform plot.

Another problem that is still a reality in Brazil concerns violence in the countryside, Barcellos points out:

> According to figures released by the Pastoral Land Commission (CPT), the number of conflicts over land rose by 42 per cent. In addition, the number of families victimised by gunmen rose to 19,968 in 2012. An increase of approximately 30 per cent, the highest rate since 2004 (2013, p. 01).

Throughout Brazil's agrarian history, latifundia have been strengthened by countless processes inherited from the colonial period. On the other hand, the struggle to gain access to land is evident in many parts of Brazil.

Analysing the plans and projects relating to the Brazilian agrarian question, we can see that there is a discrepancy in their applicability. Analysing the distribution of land in the country, we can see that latifundia still reigns supreme, encouraged every day by the current production model. Access to land is restricted for a wide range of Brazilians, who wait idly for rural settlements to be set up, if not in the midst of conflicts in the struggle for land.

These plans and projects need to be put into practice, not just in political speeches or merely written down in the form of laws and decrees, but so that they actually result in the development of rural areas in the country.

There have been great achievements on the part of the Rural Social Movements, with greater investment in public policies for the countryside, intensifying the number of rural settlements, investments and aid through land credits, among others. However, despite this progress, the agrarian situation in Brazil is still precarious, and there is a need to increase, accelerate and finalise projects aimed at the countryside, strengthening small and medium-sized rural properties, avoiding the emptying and abandonment of properties, as well as ensuring a dignified life in the rural environment.

3.1- Considerations on Agrarian Reform Programmes in Brazil.

When dealing with issues relating to the struggle to obtain land in Brazil, it is **important to understand the term "Agrarian Reform". Law** No. 4.504 in Art. 1, & 1 (BRASIL, Legislação..., 1964) addresses the term considering Agrarian Reform **as " [...]** the set of measures aimed at promoting a better distribution of land, through changes in the system of its possession and use, in order to meet the principles of social justice and increased productivity."

Political debates on the need for solutions to the **problem of Brazil's agrarian**

scenario are recent, **knowing that "[...] there has been debate over a** period of only 60 years, which is very little in relation to the development of **our society." (STEDILE, 2005, p. 12).**

In Brazil, the demands for agrarian reform were born in the 1950s, initially marked by a humanitarian ideology aimed at resolving injustices stemming from the early colonisation of Brazil, since its formation was based on the exploitation of the land. With the arrival of the Portuguese in 1500 on Brazilian soil, there was a considerable change in the dynamics experienced by the natives, as the search for products with commercial value triggered a new rhythm **for the "new land"** (emphasis added) that had been discovered.

Thus, in **this process of invasion, "they managed to dominate the entire territory and subject the people who lived here to their way of production"** (STEDILE, 2005, p. 19). Production destined for export was encouraged by the **"concession of use" which gave farmers the right to leave large** properties as inheritance, so the land could not be sold, but it gave them the right to continue production in the same family, which was profitable for the Portuguese crown as well as for the large producers.

During the colonial period, Brazil had an agro-export model, at which point the country's development was delayed, and it was even one of the last countries to adhere to the abolition of slavery.

After the abolition of slavery in Brazil and due to the export crisis, the first models of the peasantry were formed in the Brazilian population, happening in two ways, the first with the migration of poor European peasants, and the other with the emergence of the Brazilian sertanejo peasant, who was made up of mestizos:

> [...] excluded by the land law of 1850 from the possibility of becoming small landowners, they then began to enter the "sertão" in the more inland regions of Minas Gerais and the whole of the Brazilian Northeast, in search of public lands that would not be disputed by the capitalist producers, concerned with producing for export and who occupied the best lands located on the coast and close to the ports. (STEDILE, 2005, p. 13).

The first time it was mentioned that Brazil had a serious agrarian problem was in 1946 by the Communist Party of Brazil (PCB), which pointed to the accumulation and concentration of land ownership since 1850, caused by the Land Law, as the main factor. The concentration of land, inherited from the beginning, strengthened the need to fight for agrarian reform:

> The demand for agrarian reform, likewise, was born in the 1950s as a demand from enlightened sectors of the urban middle class, from conservative and family-orientated Catholic sectors, marked by moderate and cautious commitment, from some left-wing Catholic sectors and from a fraction of the secular left. [...] this "out-of-place" origin in the middle class has given the struggle for agrarian reform an intense ambiguity, from which it has not been freed to this day. It's enough to bear in mind that under the same Agrarian Reform label there were disparate projects to

intervene in property rights, always on behalf of third parties, the rural workers. Groups that were more than antagonistic, they were clear enemies (MARTINS, 2000, p. 102).

An Agrarian Reform Programme was presented by the PCB in this period by then Senator Luiz Carlos Prestes, with the main aim of alleviating the problematic situation in the Brazilian countryside at that time. Table 2 below shows Senator Prestes' speech on the situation in rural Brazil to the National Constituent Assembly in 1946:

Chart 02: Situation of Rural Brazil Presented by Senator Prestes in a Speech to the National Constituent Assembly in 1946.

Of Brazil's 41,574,894 inhabitants, 28,432,831 live in the countryside, a total of 68.39 per cent.
Of these, 9,166,825 make up the working population (over 10 years old), people directly linked to agricultural production, who represent 67.40 per cent of the entire working population in Brazil and a further 32.24 per cent of its rural population.
Of those who work in agriculture and livestock, there are only 1,903,868 rural properties, which means that only 20.8 per cent of those who work in agriculture and livestock, or 6.7 per cent of rural dwellers, or 4.6 per cent of Brazil's inhabitants, own them.
The total area of farms, 197,626,914 hectares, represents only 23.2 per cent, so much of it is still unpopulated.
The cultivated area in Brazil, 12,921 thousand hectares, is no more than 6.5 per cent of the total area of rural properties, or 1.5 per cent of the Brazilian territory. This means that most of them remain unexploited, constituting veritable latifundia.
The cultivation of maize, coffee and cotton (the last two for export) represents 56 per cent of the entire area cultivated in Brazil, and if beans, rice, manioc and sugar cane are included, it rises to 90 per cent. This means that our agrarian economy rests on the extensive exploitation of a few products, two of which are destined for export, and is currently in crisis.

SOURCE: Stedile, 2005, Organised by SANTOS (2013)

For Prestes, this backwardness is the traditional model of hoe agriculture similar to that practised in Egypt during the time of the pharaohs, which is difficult to get out of because the use of technology is impossible due to the millions of labourers out of work. This can explain the existence of a large mass of Brazilian peasants who are excluded by the modernised mode of production developed by the large establishments, which in turn have the technological apparatus needed for large-scale production, while peasants are left to continue with the traditional model of cultivating the land, i.e. the hoe.

Still in his speech, Prestes mentioned that farmers, in the name of agriculture, sought credit from the Bank of Brazil, however, this credit was destined for the coffee industry, and the money was used in sectors other than agriculture. For Prestes, agrarian credit was indispensable in Brazil, and those who really wanted to cultivate the land were penalised by the big producers. For him, the landowners found it easier to obtain the credit available for agriculture from the Bank of Brazil.

From 1945 onwards, many other projects were presented to improve the Brazilian agrarian question, always focussing on a fair distribution of land for agricultural production, giving landless rural workers this right, even if it was considered utopian by many. Agrarian reform measures were presented over the years, such as the Agrarian Reform Bill presented by deputy Leonel Brizola in 1963, and yet another in the government of João Goulart in 1964. Table 3 below lists some of the prominent public actions in 1962, 1963 and 1964.

TABLE 03 - Public Actions Aimed at Brazil's Agrarian Issues in 1962, 1963 and 1964.

YEAR	PUBLIC ACTIONS	OBJECTIVE
1962	Creation of the Superintendence of Agrarian Policy (SUPRA)	Carry out agrarian reform in the country.
1963	Rural Labour Statute	Regulating labour relations in the countryside.
1964	Land Statute	First National Land Reform Plan .

SOURCE: SILVA, A. (2009) Organised by SANTOS (2014).

However, for Silva (2009, p. 20), the programmes created by the government at this time "only served to make up for the agrarian question, as the number of families expected to be settled was much lower than planned".

The Land Statute was seen as progressive legislation during the period of the military government, instituted for the first time in the country. Table 04 below shows the functions and elaborations of the Land Statute.

TABLE 04: Functions and elaborations of the Land Statute.

Registration of all land properties in the country
Creation of IBRA (Brazilian Institute of Agrarian Reform), responsible for registering properties and colonising public lands. Later replaced by INCRA (National Institute for Colonisation and Agrarian Reform).
Creation of the Institute for the expropriation by the state of properties that underutilised their productive potential.
A general classification for all properties, based on size, utilisation and production capacity, according to the size of the property it would be considered a small estate or a large estate.
The expropriation of land for agrarian reform purposes from properties classified as smallholdings, in order to regroup the area; and from large estates, with the aim of land distribution.
It makes it compulsory to pay ITR (Rural Land Tax), which until then had not existed, and the funds are earmarked for the Agrarian Reform Programme.

SOURCE: Sdetile (2005), Organised by SANTOS (2014).

Another important moment in Brazil's agrarian history was the creation of INCRA (National Institute for Colonisation and Agrarian Reform) through Decree No. 1.110, of 9 July 1970, with the aim of calming the large mass of people demanding agrarian reform policies, as well as administering the Union's public lands, with a total of 30 regional superintendencies.

Through the concept of territorial development in order to establish rural settlements

38

with the aim of utilising the compatible potential of each biome, in this sense, the peasant economy would present positive results, it is interesting for the small rural producer to know about the area destined for the rural settlement, so that he can exploit it in a positive way.

the soil in order to produce goods with greater security, guaranteeing income for the family group, which implies the consolidation of the property.

As far as Mato Grosso is concerned, during the military period, INCRA was in charge of 60% of the state's land in the process of land regularisation and occupation of Mato Grosso's territory. **INCRA was responsible for "discriminating and** collecting the vacant lands existing there, as well as deciding on their destination, in compliance with the **Land** Statute [...]" (MORENO, 2007, p. 156).**

During the military dictatorship, the Federal Private Colonisation Programme was created, divided into Business Colonisation and Settlement Colonisation, which respectively had the objectives of proposing the establishment of agricultural, agro-industrial and agro-mineral projects, among others, and connecting land policy to economic policy, favouring the installation of large-scale capitalist companies in border areas:

> [...] part of the geopolitical strategy of occupation/exploitation of the Amazon, for its rearticulation into the national and international economic-political system, according to the policy adopted for the country by the military governments. Considered the "gateway to the Amazon", Mato Grosso became part of capitalism's extensive development process and was awarded a plethora of special development programmes [...], which primarily served to sponsor access to land in the region by large economic groups (MORENO, 2007, p. 156).

Table 05 below lists the programmes developed during the military dictatorship, as well as the year and the meaning of the acronyms.

TABLE 05: Programmes Developed during the Military Dictatorship.

PROJECT	YEAR	MEANING OF ACRONYMS
PIN	1970	National Integration Project
PROTERRA	1971	Programme to Redistribute Land and Stimulate Agroindustry in the North and Northeast
PROVALE	1972	Special Programme for the São Paulo Valley Francisco
POLAMAZONIA	1974	Amazon Agricultural and Agromineral Hubs Programme
POLONORDESTE	1974	Programme for the Development of Integrated Areas in the Northeast

SOURCE: Moreira, M. (2008), Organised by SANTOS (2014).

The struggle for land that exists in the country today is, in general, another chapter in the history of the Brazilian peasantry. There are considerable new developments in this struggle today, starting with the process of (re)peasantisation of landless families through their settlement on rural property.

Rural settlements have become the site of a **complex and sophisticated process of (re)construction of "peasant territory" (emphasis added)**. For Simonetti (1999), this is a solid demonstration of the territorialisation of the movement, i.e. the struggle for land, not just a place of production, but also a place where life is built. According to Covezzi apud Carvalho (2012, p. 28), a rural settlement is:

> [...] it consists of an area made available by INCRA, in conjunction with grassroots social movements and, more recently, with the municipality's Municipal Council for Rural Development (CMDR), where families selected by INCRA in camps have been placed, based on their aptitudes, distributed in plots that theoretically have economic viability.

The development of projects that focus on improving the Brazilian countryside and their applicability provides support in the fight against the concentration of land in the territory. In short, from the very beginning of the colonisation of Brazil, the process of concentration and exploitation of the soil was attributed to the countryside. This contradictory and exclusionary scenario has led to the development of ideas, projects and plans that have favoured small landowners as well as those who do not yet own land.

The struggle for the right to land, together with the application of measures related to the rural environment, contributed to the process of deconcentration of Brazilian land, directing, albeit slowly, other directions in the country's agrarian relations. From the 1980s onwards, countless social demonstrations began, including the creation of the rural social movement, which gave rise to the Landless Workers' Movement (MST). It played an important role in the struggle for the right to land, with the MST's **1984** Platform of General Objectives **considering "landless" people to be "landless":**

> Rural workers who work the land under the following conditions: partners, sharecroppers, tenants, aggregates, sharecroppers, squatters, occupants, permanent and temporary wage earners and smallholders with less than 5 hectares (STEDILE, 2005, p. 177).

The role of social movements in favour of agrarian reform has been and still is of the utmost importance. Since the 1990s, with increasing criticism of the capitalist industrial agricultural model and its negative effects on the planet's natural and cultural environment, the peasantry has been given a political foothold. Thus:

> [...] the Landless Rural Workers' Movement (MST) began to define itself as a

The MST fights for immediate agrarian reform, for social equality in the fight against capitalism, defending the idea that the land should belong to those who work it. The movement also seeks the support of other sectors of society such as trade unions, churches and other organisations, always publicising the struggles and victories won by the MST.

In this way, the MST promotes the struggle for land, the challenge of conquering the right to property, guaranteeing the permanence of the peasant way in Brazil, settling countless families deprived of access to land, excluded by the capitalist mode of production, the latifundia and export agribusiness, which increasingly aggravates this struggle between large landowners and landless rural workers.

Issues related to "Agrarian Reform" give rise to heated discussions, with proposals that, off paper, would bring about progress for the Brazilian population, but the discourse retracts into the chambers, and much of what is drawn up does not reach the masses who are the focus of the discussions - those who **own small properties or even the idle so-called "landless".**

In its 2002 election campaign, the Workers' Party (PT) **presented an agrarian programme with the following theme: "Dignified life in the countryside",** adhering to proposals for sustainable and solidarity-based development policies. Under the Lula government, the intention to invest in the countryside reached even **greater heights, because in the rural environment "[...]** primary, secondary and tertiary **economic activities are developed.** The combination of these sectors [...] leads to a **type of economic growth (STEDILE, 2005, p. 211, vol. 03).**

Based on a discourse of fair land distribution, guaranteeing food supply for the Brazilian population, recovering natural resources, among other proposals by the Lula government, the issue of agrarian reform is still an important topic for political debate, as it is clear that the support of public policies is necessary for the programmes to be actually implemented as presented, making the state's supporting role a reality. Table 06 below shows the seven axes of the Lula government's Agrarian Programme:

TABLE 06: Seven Strategic Axes of the Lula Government's Agrarian Programme.

AXIS	BASIC POLICIES
Promotion and	- Sovereign and qualified insertion of agriculture in the internal and external markets;

defence of national agriculture	- Creation or strengthening of mechanisms to protect our agriculture from unfair competition from imports or speculative price fluctuations on the international market; - Implementation of mechanisms that result in income protection for farming families; - Encourage more balanced relations between farmers, agro-industries, distributors and consumers.
Strengthening family farming	- It advocates strengthening family farming through greater capacity for self-consumption and market production; - Stimulating the capacity to add value to family farming products and strategies for organising production; - Promote intensive liaison with state and local authorities to strengthen family farming; - Monitoring the use of resources earmarked for family farming.
Implement a national agrarian reform policy	- Implementation of a broad, not atomised, agrarian reform programme; - Generating jobs in the countryside, contributing to food sovereignty policies, fighting poverty by consolidating family farming; - Measures to extend access to current small landowners and their children.
Generating income and quality jobs	- Maintenance of current agricultural jobs, plus the generation of new jobs with the Agrarian Reform; - Incentives for the densification of agro-industrial chains in different Brazilian regions; - Broad access to social policies such as: health, education, housing, infrastructure, among others, as well as the creation of new jobs.
Building citizenship in rural areas	- It is essential to link emergency actions with structural actions, breaking the false dichotomy between the economic and the social; - Investment in generating opportunities and making available technologies that are compatible with the reality of the countryside; - Formulation and implementation of policies aimed at this social segment, with a participatory and mobilising nature.
Food sovereignty and security	- Quality food is the inalienable right of every citizen, and it is the duty of the state to create the conditions for this; - The right combination of structural policies aimed at income redistribution, production growth, job creation, land reform, among others; - Working together with: organised civil society, trade unions, associations, NGOs, universities, schools, churches, business organisations, etc.
Building territorial policies for sustainable	- Rural development needs to be part of a regional development policy; - Articulate the previous axes by overcoming the current segmentation, promoting new links between urban and rural, agricultural and non-agricultural;

| development | - Regional planning and regional policies as fundamental bases for other development policies. |

SOURCE: Stedile (2005), Organised by SANTOS (2014).

In the government of Luiz Inácio Lula da Silva, the PNRA was reformulated and presented as the II PNRA. According to the MDA's website, the aim was to guarantee land to 530,000 families by 2006, with 400,000 families settled through the agrarian reform programme and another 130,000 through the National Land Credit Programme.

According to the MDA (Ministry of Agrarian Development) portal:

> The new PNRA works with the concept of territorial development. The aim is to put an end to the idea of a single settlement model to be adopted throughout the country, but rather to establish and develop settlements according to the potential and characteristics of each region.

For the MDA, the settlements of landless labourers are part of the agrarian reform, guaranteeing the progress of the projects and plans already approved.

Among the Lula government's policies for sustainable rural development and solidarity is the PNRA, which was the start of a concrete agrarian reform programme. With the PNRA, rural settlements have been set up throughout Brazil, seeking to meet the needs of those waiting to obtain land. However, there is still a large number of Brazilians waiting to be given plots of land for the so-called agrarian reform, which means that there is a need to intensify the development of projects relating to the countryside as well as their realisation. Table 07 below shows the operational objectives of the Agrarian Reform Programme:

TABLE 07: Operational Objectives of the Agrarian Reform Programme.

Promote the establishment of reformed zones, prioritising the expropriation of unproductive land.
Make the programme viable through the use of TDAs (Agrarian Debt Bonds), with measures to reduce compensation costs.
Guaranteeing human rights by promoting permanent rural labour inspections and combating violence in the countryside.
Guaranteeing social and economic infrastructure, technical assistance, access to rural credit and marketing policies, in partnership with the state and municipalities.
Drawing up settlement plans in line with environmental preservation.
Developing specific actions for indigenous and quilombola communities, as well as the demarcation and regularisation of their lands.
Confiscate properties that practice slave labour for the purposes of agrarian reform, in accordance with the law.

SOURCE: Stedile (2005) Organised by SANTOS (2014).

As has been noted, government policies have been disseminated since the 1970s and are based on combating social inequalities. Even so, the percentage of landless families and others without the aid of public policies to encourage them to remain in the countryside is extremely high in Brazil, and there is a need to speed up projects and plans in favour of

Agrarian Reform.

In order to develop truly effective land reform programmes in the country, two structural elements need to be considered. The first is to recognise that land is the basis for human existence and its use in this sense is to satisfy the essential needs of life for human consumption. The second would be to recognise that peasant agriculture allows the social function of land to be established. With this recognition, it would be possible to minimise the problems related to the agrarian question in Brazil.

4.2-The   Case of Mato Grosso

The programmes related to rural settlements played an important role in the search for agrarian reform that included those who did not yet own land. The PEA (Special Settlement Project) was designed to settle families who had left regions of agrarian conflict. In Mato Grosso, the PEA - Lucas do Rio Verde was set up on the banks of the BR-163 highway in the municipality of Diamantino, with 252 families settled from the municipality of Ronda Alta in Rio Grande Sul (MORENO, 2007).

Of the 250 families settled in 1981, with around 1,000 people in an area of 200,000 hectares, this project had national repercussions due to the scandals and corruption related to the implementation of the PEA involving INCRA officials, according to Moreno (2007), of the 252 families settled, only 15 remained, and in 1982 there was an improvement of the Project with the participation of the Lucas de Rio Verde Cooperative (Cooperlucas) and 972 more families were settled there.

In the first wave, many of those who left the settlement sold their plots to the settlers who remained, and with this "large properties were formed and Lucas do Rio Verde became one of the municipalities with the highest agricultural production in the state, based on the monoculture of soya and cotton". (MOREIRA, 2008, p. 37).

In Mato Grosso, between 1968 and 1992, 33 colonisation projects were set up, among which the ones that stood out the most were those in the region **popularly known as the "Nortão"** (our emphasis) of the state, including the cities of Sorriso and Sinop, with populations mainly from the south of Brazil, the result of the federal government's policy of encouraging the occupation of central Brazil, partly solving the problem of the many families who were fighting for the right to land.

The emergence of municipalities considered agricultural is part of the state's reality, especially in the north. From 1948 onwards, there was a large influx of migrants from the southern regions, creating a new cultural expression, especially in the south.

The process of occupying the territory reached greater proportions from the 1970s onwards, **with the insertion of the agricultural frontier, intensifying the "[...]** migratory **flow**, fostering the creation of new municipalities, consequently new cities, as well as increasing the **growth of existing cities."** (VILARINHO NETO, 2009, p. 13). This process of occupying the so-called empty spaces was an attempt to momentarily remedy the conflicts that had arisen over obtaining land in the north-east and south of Brazil.

The concentration of large estates in Brazil is a reality, and in the case of Mato Grosso this situation extends even further. As Moreira (2012, p. 47) points out:

> While in Brazil agricultural establishments of up to 10 hectares occupied an area of 2.36% of the Brazilian agricultural area, in Mato Grosso this group occupied just 0.12%. Establishments in the group of more than 1,000 hectares, which in Brazil occupied 44.42 per cent of the area, in Mato Grosso covered 77.51 per cent of the agricultural area.

This data indicates that the state of Mato Grosso needs greater attention from the Brazilian government in favour of fair and effective land reform. As the state with the highest rate of soya production, there is also a tendency to expropriate those who are not integrated into the large-scale production system.

The scenario of land concentration in the state of Mato Grosso today is a reflection of a past of informal occupation, because **"in the early** days of the occupation of Mato Grosso, land ownership was by occupation, because there were no laws regulating the occupation of public lands, it was only in 1850 **that Land Law No. 601 came into being". (CARVALHO, 2012, p. 57).**

The Land Law was applied in the state for the first time in 1892, in order to regularise the land situation in Mato Grosso and in the same year the distribution of public land was also supported by law, and the large estates in Mato Grosso were guaranteed large portions of land, even those that did not comply with the Land Law of 1850 because they had an area larger than 3,600 hectares were able to regularise their property through the state law (LAMERA and FIGUEREDO, 2007).

The formation of agriculture took place after the decline of the mining process in the state, because during the gold-mining period (18th century), cultivation was aimed at subsistence, with gold extraction considered to be of greater economic importance. Thus:

> The fall in production, combined with the low quality of alluvial gold and high taxes, plus the discovery of new deposits in the region, caused a period of decline in gold mining. After this period of stagnation, almost a century after it was founded, Cuiabá was granted city status by Royal Charter in 1818 and declared the capital of Mato Grosso in 1835, seventeen years after it was founded. (CARVALHO, 2012, p. 59).

From 1992 onwards, 59 settlement colonisation projects were developed in different locations in Mato Grosso, classified according to the implementation strategy adopted. The PAR (Rapid Settlement Project) was aimed at locations that already had a certain infrastructure, with plots averaging 50 hectares. 9 PARs were set up in the state, serving 4,542 families in the municipalities of Colíder; Nova Canaã; Pontes e Lacerda; Sto. Antonio do Leverger; Aripuanã. On the other hand, the PAC (Joint Action Project) is a joint action between INCRA and the cooperative, with 03 projects, settling 7,459 families in the municipalities of Nobres, Alta Floresta and Guarantã do Norte (MORENO, 2007).

Another project aimed at rural settlements is the PA (Settlement Project), initially designed for areas of conflict with squatters who were already encamped. INCRA is responsible for carrying out the process to consolidate the projects. In this modality, the state had 46 projects between 1981 and 1992, totalling 12,830 settlers, and it was later continued through the PRRA (Regional Agrarian Reform Plan). (MORENO, 2007).

Faced with the exclusionary process of the capitalist mode of production and the strengthening of the latifundia, the struggle for land ownership became viable, with the presence of organised social movements gaining a solid foundation in the demands for land rights. The first clashes involving the MST took place in 1986:

> [...] in the municipality of Lucas do Rio Verde in the mid-north of the state, on the banks of the BR-163 motorway, which in turn differs from Sorriso and Campo Novo do Parecis in terms of its formation/creation, as it is a rural settlement promoted by the federal government in the 1980s, on land belonging to the union. [...]As a result of this settlement project, the municipality of Lucas do Rio Verde was formed, which, because it is also in a region of arable land, was another phenomenon of urban expansion. Today, with approximately twenty-five thousand inhabitants, the city has a commercial centre that meets the needs of its residents, both in the domestic and rural and industrial areas. (CARVALHO, 2012, p. 64).

Another interesting fact is the location of the settlements, given that the large properties are located close to the highways, allowing production to be transported, the vast majority of which is monoculture crops such as soya, corn, cotton, etc. **As MOREIRA (2008, p. 48) mentions,** "areas far from the road infrastructure have seen the development of the federal government's rural settlement policy." Table 08 below shows the distribution of settlements in the territory of Mato Grosso.

TABLE 08: Distribution of Rural Settlement Projects by Mesoregion in Mato Grosso:

MESORREGION	MUNICIPALITIES
South East	Rondonópolis; São Pedro da Cipa; Dom Aquino; Itiquira; Jaciara; Juscimeira; Pedra Preta; São José do Povo; Alta Araguaia; Campo Verde; Araguainha; General Carneiro; Guiratinga and Poxoréo.
North-East	Agua Boa; Alto Boa Vista; Nova Nazaré; Nova Xavantina; Novo São Joaquim; Querência; Campinápolis; Araguaiana; Barra do Garças; Cocalinho; Canabrava do Norte; Confresa; Novo Sto. Antonio; Porto

	Alegre do Norte; Ribeirão Cascalheira; Santa Terezinha; São Félix do Araguaia; São José do Xingu, Serra Nova Dourada; Vila Rica.
North	Alta Floresta; Nova Monte Verde; Apiacás; Carlinda; Nova Bandeirante; Paranaíta; Ipiranga do Norte; Nobres; Nova Mutum; Nova Ubiratã; Nova Maringá; Sta. Rita do Trivelato; Sorriso; Tapurah; São José do Rio Claro; Tabaporã; Aripuanã; Brasnorte; Castanheira; Colniza; Cotriguaçu; Juina; Juruena; Colíder; Guarantã do Norte; Matupá; Nova Canaã do Norte; Nova Guarita; Novo Mundo; Peixoto de Azevedo; Terra nova do Norte; Gaúcha do Norte; Paranatinga; Planalto da Serra; Comodoro; Diamantino; Claúdia; Feliz Natal; Marcelândia; Sta. Carmem; União do Sul; Vera and Nova Brasilândia.
South West	Pontes e Lacerda; Vila Bela da Santíssima Trindade; Nova Lacerda; Araputanga; Jauru; Mirassol d'Oeste; Figueirópolis d'Oeste; Porto Esperidião; Reserva Cabaçal; Rio Branco; São José dos Quatro
	Marcos; Barra do Bugres; Denise; Nova Olímpia and Tangará da Serra.
South Centre	Barão de Melgaço; Cáceres; Curvelândia; Poconé; Alto Paraguai; Nortelândia; Nova Marilândia; Santo Afonso; Chapada dos Guimarães; Cuiabá; Nossa Senhora do Livramento; Santo Antônio do Leverger; Várzea Grande; Acorizal; Jangada and Rosário Oeste.

SOURCE: INCRA - SIPRA Report (2012).
Organised by: SANTOS (2014)

As noted in the table above, the Pantanal municipalities of Mato Grosso have been the target of Brazilian land reform. The distribution of settlements in the region requires government support because it is a complex area, different from the other regions of Mato Grosso.

4.2.1-Comparison between rural settlements in different regions of Mato Grosso: the cases of the Igarapé do Bruno, 14 de Agosto, Corixinha and Santo Onofre settlements - MT.

As we have already seen, rural settlements are spread across all of Mato Grosso's mesoregions. Work related to rural settlements in Mato Grosso is significant for understanding agrarian reform in the state, and also provides support for understanding the relationships that exist in rural areas in its different regions. In order to better analyse and present the results obtained in the study area, namely the Santo Onofre Settlement, a comparative presentation was constructed between rural settlements in different locations in the state.

The practice of Settlement Projects (PA) takes place in two phases, the first being implementation and the other consolidation. The first phase is dedicated to selecting the families destined to be settled, going through a pre-registration process followed by an interview, after which comes the legitimisation and homologation process (SILVA, 2009).

The implementation is carried out through the contemplation of the settlement contract and a Settlement Development Plan (PDA) is drawn up afterwards. The settlers are given access to training and instructions on how to work with the land by specialised technicians. Topographical work is also carried out to establish the perimeter of each plot in the area, basic infrastructure is put in place to accommodate the families, and another important factor is the support of credits such as PRONAF to subsidise activities on the property (SILVA, 2009). The PA is consolidated once all the implementation phases have been completed.

Silva (2008) carried out a study of occupation and socio-environmental conflicts in the Igarapé do Bruno settlement, located in the municipality of Apiacás in the far north of Mato Grosso. The settlement was granted an area of 9.8 thousand hectares, covering 237 families, of which 132 families currently remain.

Located in a region dominated by the Amazon, the soil has low fertility, the agriculture practised in the settlement is basically subsistence, and a large part of the area is used for livestock, beef and dairy farming (SILVA, 2008). Access to the settlement is via five unpaved roads. Access to towns is an important factor for transporting produce, but some of the roads are not in good condition, which is an obstacle for rural dwellers.

Still with regard to the social adversities in Igarapé do Bruno, the main ones cited are related to the growth of large estates in the area. This is explained by the fact that there are **"latifundia** formed through the acquisition of land from estate farmers who left the countryside for the city or other regions." (SILVA, 2008, p. 58). Another factor pointed out is the lack of opportunities to diversify production, causing the property to be less utilised.

Accessory labour is a reality among families in the area, and the search for parallel income is necessary for domestic supplies and sustenance. The recommendation made by Silva (2008, p. 62) with regard to the Igarapé do Bruno settlement is that:

> [...] it's not enough to settle rural workers and distribute land. More is needed. Today, production units are required to produce enough, generating surpluses for their owners, so that they can modernise their farming activities from their profits. It's also important that everyone has access to good infrastructure services such as electricity, roads, transport for production and, of course, credit lines.

Another study related to rural areas concerns the 14 de Agosto settlement in the municipality of Campo Verde. Silva (2009) researched the economic and social characteristics of the settlement. The settlement has an area of 1,750 hectares divided into 70 plots, of which 28 have already been sold.

The first activities developed in the settlement were related to livestock farming, because there was a cattle ranch in the area with pastureland, which made it easier to raise cattle. The lack of studies related to agricultural suitability made commercialisation difficult at first, and the crops planted were according to the wishes of each producer.

The settlement has a health centre and a cooperative with 12 member families. The area set aside for the cooperative is collective and includes a tractor, lorry, canteen, tank, flour mill and other facilities. According to Silva (2009), the settlement's production base is related to sugar cane derivatives; manioc, flour and starch production; vegetables; milk, as well as subsistence crops such as rice, beans, coffee, eggs and green maize.

The products are marketed among local residents and at the Campo Verde municipal market. They are also sold in the settlement itself to buyers in local markets, as well as to Cuiabá and Rondonópolis. Sugarcane by-products such as brown sugar and burning water are traded via CONAB (National Supply Company).

The development of the 14 de Agosto settlement (Silva, 2009) is related to the following factors: settlers of peasant origin; strategic location and consumer market; access to credit lines; the presence of a cooperative; the practice of pluriactivity and suitable physical conditions. Thus:

> A vitally important factor for the settlers' economy is their vast consumer market, which meets local demand as well as that of the state's main commercial centres. In the case of fruit and vegetables, in some cases the buyers come to the settlement to pick up the product. [...] In terms of public policies

> (SILVA, J., 2009, p. 43 - 44).

Another example is the case of the Corixinha Settlement, located in the municipality of Cáceres in the Pantanal region of Mato Grosso. The research carried out by Moreira (2008) addressed the issue of family farming and public policies, in this case the implementation of PRONAF in the Corixinha Settlement.

The settlement was earmarked, as Moreira (2008, p. 80) describes, for the purpose of setting up rural establishments by decree:

> The decree declaring Fazenda São Judas Tadeu/Corichinho to be an area of social interest for land reform purposes was published in the Federal Official Gazette on 17/09/1998, with a total area of 3,413.1808ha [...]But it was only on 05/04/2001, by means of decree No. 006/05, that INCRA - MT declared the Corichinho Settlement Project to be established, for the purpose of settling 72 families.

In the area reserved for settling families, 8 hectares were set aside for the construction of churches, a school, the association's headquarters and a small shop. The main house, built by the residents themselves, has a thatched roof. The average number of family members is between three and five, all of whom take part in the activities carried out on the property (MOREIRA, 2008).

With regard to the permanence of landowners on the plot, there was a high percentage of sales of ownership rights to third parties. This process is a consequence of the lack of public policy support for the area. The main difficulties pointed out are: "[...]

isolation, the lack of quality education and schools, health care structures, technical assistance and means of commercialising production." (MOREIRA, 2008, p. 87).

A charcoal factory has been set up in the settlement which, according to the owner, is the only source of income in the area. Among the problems and difficulties that exist in the region are the poor condition of the school and the bus that transports the students, better health care, lack of water and tractors to work the land. Another serious problem relates to access to PRONAF:

> When it comes to families excluded from PRONAF, they are second or third owners. Officially for INCRA, these families are not part of the PNRA and are therefore ineligible to access credit. Those who are PRONAF beneficiaries have difficulty detailing the conditions under which they borrowed the money. There is a lack of information about the grace period, the instalments to be paid, as well as the amounts spent on each activity on the property (MOREIRA, M., 2008, p. 91).

According to Moreira, the families of the Corixinha settlement produce for subsistence, commercialising the surplus and the milk that comes from the local livestock activity. The increase in chicken and pig farming is important for the residents' food security. This whole process took place after access to PRONAF, but the installation of settlements in unsuitable locations by INCRA becomes a problem in terms of compliance with the PDA.

A comparative analysis of the aforementioned settlements and the study area of this dissertation, the Santo Onofre Settlement, reveals some similarities in the results, such as the lack of public policies that provide producers with better conditions for cultivating the land and access to commercialised goods. Another similarity is related to precarious basic structures such as schools, roads, leisure areas, agricultural machinery and implements, among others.

The agrarian dynamics of the settlements mentioned are similar when it comes to the practice of accessory labour. In the case of Santo Onofre, it's common for families to look for salaried income to meet their domestic needs; local residents consider it better to have an income with a guaranteed monthly amount than to wait for the land to produce without being sure that it will be commercialised.

That's why cooperatives are so important, as in the case of the 14 de Agosto settlement, where the population has a guaranteed market and can work the land safely. However, not everyone has such a facility, making the situation even more difficult for small rural producers, such as the inhabitants of Santo Onofre.

The process of selling off plots is a common occurrence in rural settlements, and all those mentioned had the same situation with regard to families returning to the city. The number of abandoned plots or plots with residents who have bought the right of possession is considered to be high, and this reality is also related to Santo Onofre. Of the 95 families who took possession, only 23 still live there, the rest have returned to the urban environment, mainly to the cities of Poconé, Várzea Grande and Cuiabá, while others have opted to go to the property only at weekends, preferring to continue living in the city, not to mention the completely abandoned plots.In the Santo Onofre settlement, the practice of polyculture and multi-cropping contributes directly to the families' livelihoods. The variety of crops produced means that the soil in the region is suitable for agriculture. Despite the difficulties inherent to life in the countryside, when asked about their level of satisfaction, the residents said that they like living in the countryside, their relationship with the land is harmonious, and even though it is hard work, they will remain in the settlement believing in better times.

CHAPTER 5

PHYSICAL CHARACTERISTICS AND SUGAR REFORM IN THE PANTANAL REGION OF MATO GROSSO.

5.1- The Pantanal Region.

As the Onofre Settlement is located in the Pantanal region of the municipality of Poconé, studies related to **understanding the theme of** "region" become viable in order to better understand the Brazilian Pantanal, its dynamics and the anthropogenic relationships that exist within it.

The concept of region adopted in this research is related to the physical characteristics interconnected with the relationships expressed by man, both as a social, cultural and economic organisation.

For the field of Geography, "Region" is one of its main Categories of Analysis. The first proposals for the regional division of Brazil took place in 1913, for use in teaching geography in schools. In 1941, according to Vilarinho Neto (2009), with the aim of summarising and organising Brazilian regions, the IBGE (Brazilian Institute of Geography and Statistics) drew up new regional division surveys, taking into account physical and socio-economic aspects, which were adopted by the federal government in 1942.

In 1969, regional divisions took on the forms used today, with Brazil divided into the North, Northeast, Centre-West, South and Southeast regions. For Lencione (2007) apud Carvalho (2012), the region is a spatial cut-out, linked to the complex differentiation evident in today's world, suffering direct political influences, a fact evidenced in Brazil by the main regional divisions presented at the time, which have their own physical and cultural characteristics.

The author also emphasises that studies based on the concept of the region are an intense search for an understanding of aspects of social and cultural life. Therefore, as Santos apud Carvalho (2012, p. 34) mentions, the region can be considered:

> [...] in fact, the *locus of* certain functions of the total society at a given moment [...] The region would thus be defined as the result of the possibilities linked to a certain presence in it of fixed capital exercising a certain role or certain technical functions and the conditions of its economic functioning [...]. (SANTOS, 2004 apud CARVALHO, 2012, p. 34).

Vidal de La Blache can be considered one of the great forerunners of the concept of Regional Geography, developing various studies on the **subject of "region". For La Blache, as quoted by Moreira, R. (2008), the region is** seen as a concrete entity that exists on its own and can be described and delimited; nature as a whole is included in this notion of region, including man.

Moreira, R. (2008) mentions Vidal de La Blache and his works related to the framework of French Geography, which focused on the central category of the *region,* considered as a concrete entity, that is, one that exists on its own. From then on, the regional geography we know today was born.

The region in this case is something solid, but, according to Carvalho (2012) when quoting the concept put forward by Hartshorne, the region goes beyond the concrete, the visible, it is also an intellectual construction, underpinning research into the new geography or contemporary geography.

In order to better understand "region", it is necessary, according to Carvalho (2002), to experience its reality. Thus:

> It is a territorial portion defined by the common sense of a particular social group, whose permanence in a given area has been sufficient to establish very specific characteristics in its social, cultural and economic organisation. (RIBEIRO apud CARVALHO, 2012, p.11).

When analysing regions, it is necessary to think about the processes of regionalisation, which for Becker (2003) the regionalisation of social spaces is promoted by territorialisation and the fixation of the workforce in their region, in their place. For Becker, regional development appears as a political project (or political economy) where democracy, organisation and social participation are the elements that underpin it.

Recognition of the importance of potential and the accumulation of cultural values (social capital) for regions to be able to respond positively to the challenges of globalisation, building their own development models, taking advantage of opportunities and expanding their potential.

The Pantanal region (Fig. 02), the largest continuous continental flooded area on the planet, was designated a National Heritage Site by the 1988 Federal Constitution, covering 140,000 kilometres^2 (Vieira, 2002). The importance of this ecosystem influences researchers from various fields, including the rural environment.

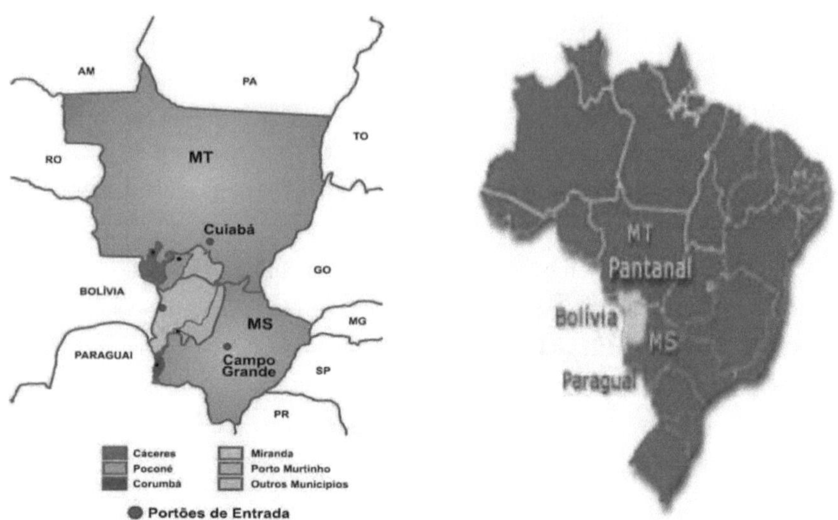

FIGURE 02: Location of the Brazilian Pantanal.
SOURCE: GOOGLE IMAGES - Organised by SANTOS (2014).

Formed by an immense floodplain located in the Paraguay River Basin, it is greatly influenced by the Paraguay River and its tributaries, which flood the region, forming extensive swampy areas (marshes) during the rainy season and the river floods between November and April.

Situated in the heart of South America, almost 90% of which belongs to Brazil, it stretches across the territories of Mato Grosso, Mato Grosso do Sul, and the countries of Paraguay and Bolivia, totalling approximately 228,000 km^2 . The Pantanal's boundaries are spread over 16 municipalities in all, as shown in figure 03 below:

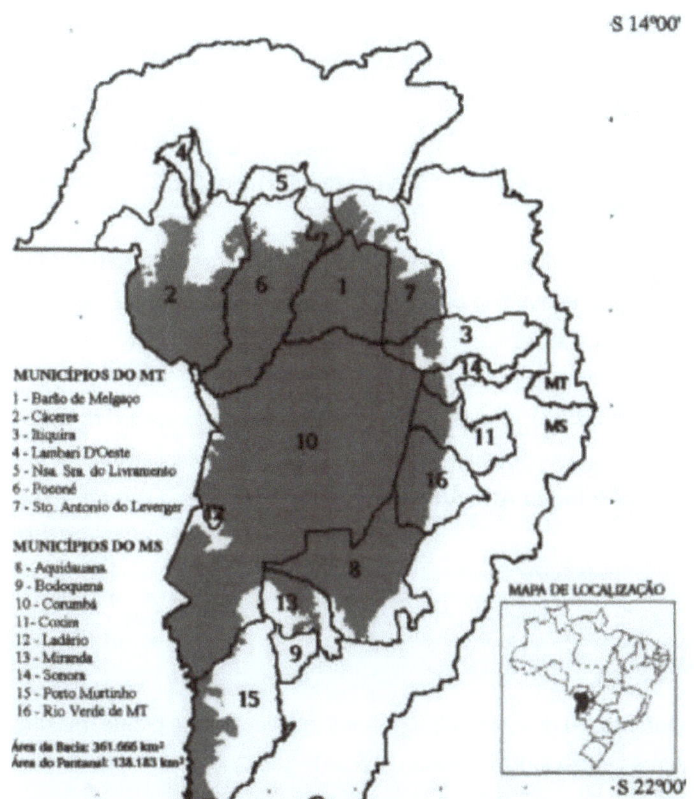

FIGURE 03: Distribution of the Brazilian Pantanal by municipality.
SOURCE: Vila da Silva and Abdon (1998).

The characterisation of the Pantanal, according to Vieira (2002, p. 01) is "[...] a sedimentary plain susceptible to periodic flooding of **varying** intensity and **duration"**. And, according to Rossetto and Brasil Junior (2001), the Pantanal is characterised by climatic seasonality, with periods of flooding as well as ebb and flow, respectively during the rainy and dry seasons. The author also emphasises that people have their own social and economic ties, giving rise to different ways of acting in this environment, the Pantanal.

Rossetto and Brasil Junior (2001) aimed to describe the relationship between the inhabitants of the Pantanal and the elements of nature, while also reflecting on sustainability in this ecosystem. When dealing with issues related to the inhabitants of the Pantanal, we know that the small producer is part of this context.

The vegetation is diverse, which is why it is called the Pantanal Complex, with plant species with characteristics adapted to the flow of water, this being the part that is periodically flooded, the part related to the plain of the Upper Paraguay Basin (BAP).

Knowing that the Pantanal has an upper part that does not suffer the dynamics of periodic flooding, as it is located in the plateau region of the BAP.

The Pantanal is made up of a wide range of animal and plant species, combining factors such as: terrestrial and aquatic habitats; low relief altimetry; geographical location; the boundary between three large natural regions such as the Cerrado, the Chaco and the Amazon.

The Pantanal's climate is classified, according to Tarifa apud Rossetto and Girardi (2012) as Tropical, with temperatures varying between 35° C and 40°C, and the minimum can reach 10° C. The altitude varies between 80 and 120 metres, and it is totally inserted in the Upper Paraguay Basin, to which it is interdependent.

The Pantanal is classified among Brazil's Morphoclimatic Domains as a Transition Area, i.e. an area that borders one vegetation domain and another, which explains the complexity of the Pantanal, **according to Aziz Ab'Saber** (apud Rossetto and Girardi, 2012, p. 139):

> [...] as a remarkable interspace of transition and contact involving strong penetrations of cerrado ecosystems; a significant participation of chaquenhas floras; inclusions of Amazonian and pre-Amazonian components; alongside aquatic and **subaquatic** ecosystems of **great extension in the "pantanais", of its great** floodplains.

It is of the utmost importance to develop public policies in favour of the biome, making it clear that the Pantanal is not just a single formation. According to Carvalho (2012, p. 25):

> The PCBAP was the forerunner of Ecological-Economic Zoning, which was developed from 1992 onwards for the national territory, with the clearly defined aim of promoting territorial planning based on environmental planning, following the principles of sustainable development. [...] the PCBAP is of great importance, both for small producers and for the environment [...].

The PCBAP (Upper Paraguay Basin Conservation Plan) is a regional planning instrument that forms part of the Pantanal Programme, coordinated by the Ministry of the Environment (MMA), with the aim of conserving this important environment. The PCBAP divided the Pantanal into 11 sub-regions (Fig. 04), according to the characteristics of the fauna, flora and physical aspects (PCBAP, 1997).

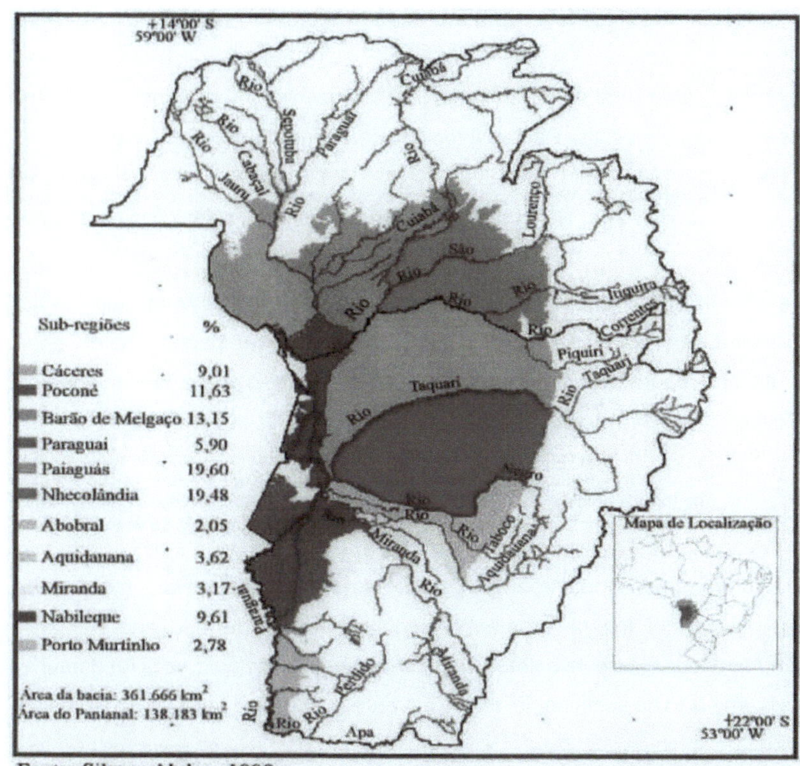

Fonte: Silva e Abdon, 1998

FIGURE 04: Delimitation of the sub-regions of the Brazilian Pantanal.

In the Pantanal, the relationships developed by man in the natural environment must take place in a sustainable and harmonious way. The activities developed in this region must obey the rhythm corresponding to the periods of drought and flood, so as not to harm this complex ecosystem, thus guaranteeing its use for future generations. Therefore, greater government assistance is needed to promote sustainable use by the Pantanal population, especially those who live in the countryside.

5. 2-Agrarian reform   in the Pantanal of Mato Grosso.

In Pantanal regions, the process of utilising the soil for agricultural production has existed for over 200 years, with the production of sugar cane and, most notably, cattle ranching, the precursor to the formation of large **properties in the Pantanal, since the "characteristics such as pastures and** natural **salt pans** made cattle ranching in the

Pantanal very profitable, which led to a **rapid increase in the herd" (MOREIRA**, M., 2008, p. 54).

The presence of large rural establishments in Pantanal regions is notorious, and this process of property extension in the Pantanal took place during the Sesmaria period:

> [...] in force in the country from 1532 and effective in Mato Grosso from 1727. This system allowed large tracts of land to be owned free of charge by people who could prove they had the financial means to exploit them (ROSSETTO and GIRARDI, 2012, p. 144).

The large size of the ranches was explained at the time by the flood pulse, which meant that more land was needed for herding cattle from the flooded parts to the dry ones. Other processes were important in clarifying the agrarian dynamics of the Pantanal region, according to Rossetto and Girardi (2012, p. 138):

> [...] the modernisation of agriculture, the dismemberment or alienation of **land through inheritance, the "closing" of the agricultural frontier in the** Amazon and the expansion of sugar cane plantations in other regions of the country, which expels livestock to other places, demanding its growth in areas such as the Pantanal.

The federal government works from the point of view of regional development policy, so the Pantanal is part of the government's planning for the Cerrado and Pantanal. Recognising the need to increase investment in research and public policies is fundamental if the use of this biome is to be sustainable, taking advantage of this ecosystem economically (STEDILE, 2005).

According to Carvalho (2012), with the agrarian reform policy in Brazil, rural settlements were set up in the Pantanal, with subsistence production from 1996 onwards, because at this point the Pantanal municipalities became targets for agrarian reform.Some government programmes were set up in the state of Mato Grosso, putting into practice the proposed development and agrarian reform plan that had a certain impact on the Pantanal, table 09 below shows the main government programmes set up in Mato Grosso from the 1960s onwards.

TABLE 09: Main Government Programmes Implemented in Mato Grosso from the 1960s onwards.

National Integration Plan (PIN)
Programme to Redistribute Land and Stimulate Agribusiness in the North and Northeast (Proterra)
Midwest Development Programme (Prodeste)
Cerrado Development Programme (Polocentro)
Special Pantanal Development Programme (Prodepan)

SOURCE: Moreno apud Rossetto and Girard (2012).
Organised by SANTOS (2014).

In Mato Grosso, according to Vila da Silva and Abdon (1998), the municipalities in Pantanal regions total an area of 48,865 km² , totalling 35.36%, with the municipality of Poconé having 80.3% of its territory as Pantanal areas, accounting for 10.21% of the total area.

There are 530 rural settlements in Mato Grosso, of which 56 are in the Pantanal region. Rural settlements in Pantanal regions began in 1997, with the presence of the MST in the municipality of Cáceres, through occupations of farms, **thus "[...]** the first settlements of rural workers in the **Pantanal** region were created by INCRA". **(MOREIRA, M., 2008, p. 55)**. Table 01 below shows the Pantanal municipalities belonging to Mato Grosso, their share of the Pantanal area and the number of rural settlements.

TABLE 01: Pantanal Municipalities in Mato Grosso, Share of the Pantanal in km² in the Municipalities, Rural Settlements by Municipality, 2012.

Territorial Unit	Area of the municipality in the Pantanal	Total area of the municipality	% of the Pantanal in the municipality	No. of settlements in the municipality
Barão de Melgaço	10.782	10.865	99,2	03
Cáceres	14.103	25.154	56,1	11
Itiquira	1.731	8.482	20,4	01
Lambari d'Oeste	272	1.711	15,9	
Our Lady of Livramento	1.115	5.134	21,7	20
Poconé	13.972	17.406	80,3	11
Santo Antonio do Leverger	6.890	11.283	61,1	10
TOTAL	**48.865**	**80.035**	**61,0**	**56**

SOURCE: Adapted from Rossetto and Girardi (2012, p. 140) and SIPRA/INCRA Report (2012). Organised by SANTOS (2014).

According to the SIPRA (2012) - **INCRA** report **on the** "Administrative **Aspects of** Property in the Process of Obtaining **Agrarian** Reform Projects" **from 1997 onwards,** the municipality of Poconé has 13, as described in table 10 below:

TABLE 10: Properties from Agrarian Reform Projects in the Municipality of Poconé.

NAME	FORM	YEAR	AREA (ha)
Furnas do Buriti	Expropriation	1997	1,061.6000
Agroana/Girau	Expropriation	1999	3,039.3000
João Ponce de Arruda	Recognition	2000	8,004.9000
Saint Philomena	Expropriation	2002	1,505.2000
Slaughterhouse	Recognition	2003	79.6000
Red Water	Recognition	2003	367.3000
Piuval	Recognition	2004	851.5000
Morro Cortado	Recognition	2004	1,008.0000
PE Capão Verde 1	Recognition	2004	225.6000
Xafaris EP	Recognition	2004	703.5000
Figueira Cologne 1	Recognition	2004	2,403.6000

and II			
Rural Village Portal	Recognition	2005	927.3000

SOURCE: SIPRA Report (2012) Organised by SANTOS (2014).

Referring to issues related to rural settlements and small rural producers, with their mode of production and survival, economic situation and difficulties faced in the rural environment, this research sought to raise these questions in the Santo Onófre Settlement, the area of our work. Although its name is not yet included in the SIPRA report due to bureaucratic issues of regularisation of documentation, it is made up of small properties and is also part of the agrarian reform.

The economy of the Pantanal region is still governed by beef cattle farming, as it is a favourable region for this practice due to the natural grassland vegetation and the possibility of extensive farming. The majority of small producers live a different economic reality: subsistence production, their relationship with the land, and are part of the Marxist M-D-M logic, where the merchandise produced is consumed and the surplus sold, which returns to the family in the form of merchandise.

The need for public policies to support production and infrastructure for small landowners, not only in the Pantanal but throughout Brazil, is essential for them to remain in the countryside, thus providing better conditions for survival in the countryside. It can be said from the results obtained in the field that the economic situation of the residents of Santo Onofre, in the face of the growth of capitalist production, is painful but persistent. The rural man elevates his feelings for the land above the economic factor, because the relationships of living overcome the obstacles imposed by the economic model in force in the country.

5.3- Brief Description of the Municipality of Poconé.

The municipality of Poconé had its origins in 1777 through the discovery of gold deposits. Its first name was Beripoconé, after the indigenous tribe of that locality (Portal Mato Grosso and its municipalities). As a result of the discovery of gold, the population began to gather more and more, and the town was founded on 21 January 1781 under the name **"São Pedro D'EL REY" in honour of King Pedro** III.

With the decline of gold in this region, the formation of livestock farms became an economic alternative, taking advantage of the natural conditions and the fertile pastureland favourable to the practice, a characteristic that still serves as the economic base for the municipality today.

On 25 October 1831, the municipality of Poconé was created and disconnected from Cuiabá, acquiring the status of a city through Provincial Law No. 1 of 1 June 1863.

According to IBGE data (2010), the municipality has 31,495 inhabitants and is part of the Cangas District, 15 kilometres from the Santo Onofre Settlement. It is located 95 kilometres from the capital Cuiabá and is accessed via the BR-070 and MT-060 motorways.

Poconé belongs to the Centre-South mesoregion of Mato Grosso, in the area known as the Alto Pantanal, with an extension of 17,261 km² , under Latitude 16°15'24" south Longitude 56°37'22" west with an altitude of around 142 m. The hydrography belongs to the Paraguay and Cuiabá Basins, both of which influence the Great La Plata Basin. The climate classification is Tropical hot and sub-humid, with relief characteristic of the highlands and Pantanal of Mato Grosso. Gold mining in the region was an important factor in the municipality's economy:

> represents an important gold district for the state of Mato Grosso and is part of a Neoproterozoic metasedimentary sequence that makes up the rocks of the Cuiabá Group. [...]. The period from 1980 to 1995 was marked by the peak of gold mining and improved control of gold production in the municipality. (LACERDA apud REZENDE, 2012, p. 20).

In addition to gold, livestock farming is also a highlight in Poconé. As a typical Pantanal municipality, the use of natural pastures encourages large-scale livestock farming, strengthening the process of land accumulation, a real characteristic throughout Brazil, but it also provides small landowners with the conditions to utilise the Pantanal vegetation in smaller-scale livestock farming.

CHAPTER 6

SOCIO-ECONOMIC ASPECTS OF THE SANTO ONOFRE SETTLEMENT

6.1- Characterisation of the Study Area.

The Santo Onofre Settlement has 95 plots of 15 hectares each, as can be seen in the sketch of the settlement (Fig. 05). The total area is 1,356.7555 ha, and the preservation area is 80.2219 ha. In 2005, the properties were drawn at a meeting with representatives from SEDRAF (Rural Development and Family Farming Secretariat); UTE (State Technical Unit); EMPAER (Mato Grosso Research and Rural Extension Company); the Poconé Agriculture Secretariat; Sindicato Rural de Poconé (Poconé Rural Union) and the settlers, who drew lots. This was adopted as a rule to allocate the location of the properties without there being any restrictions on obtaining the land, or demands to exchange it for units considered more attractive or fertile.

At first, the plots didn't have any infrastructure. According to reports from residents, there were settlers who moved into the settlement living in shacks until the houses were finished, in this case in 2007.

The area of vegetation destined for APP in the settlement is strictly respected, knowing that the Brazilian Forest Code was created in 1934 under the government of Getúlio Vargas in an attempt to order the use of natural resources, **determining "[...] the obligation to preserve sensitive areas and to** maintain a portion of native vegetation within rural properties. These are **called permanent preservation areas (APPs) and legal reserves."** **(MAPA,** 2011, p. 02).

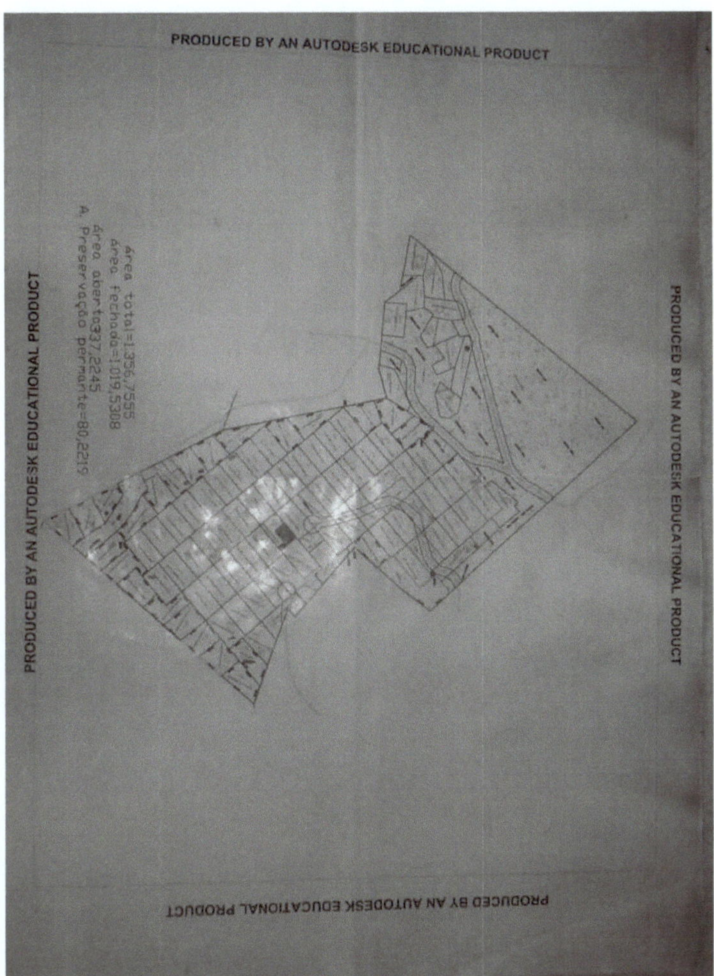

Figure 05: Sketch of the Santo Onofre settlement.
PHOTO: SANTOS (2014).

The Forest Code has been reformulated throughout its existence in 1965, 1989, 1996, 1999 and finally in 2012. With regard to the legal reserve in rural establishments, this is:]

> [...] an area located within a rural property or possession that must be maintained with its original vegetation cover. This area has the function of ensuring the sustainable economic use of natural resources and providing for the conservation and rehabilitation of ecological processes,
> promote the conservation of biodiversity, harbour and protect wild fauna and native flora. The size of the area varies according to the region where the property is located. In the Amazon, it is 80 per cent and in the Cerrado located within the Legal Amazon it is 35 per cent. In the other regions of the country, the legal reserve is 20% (Ministério da Agricultura, Pecuária e Abastecimento, 2011, p. 03).

With regard to the preservation areas in the Santo Onofre Settlement, it was noted that local residents are aware of the issue. It can be seen that the places earmarked for preservation are not suffering from degradation, practically all the plots in the settlement have a preservation area that is even larger than stipulated, which according to them, with the sustainable use of nature guarantees future generations the right to also enjoy the natural environment.

The typical vegetation in this location is the Cerrado, which is the second largest biome in South America with a total area of 2,036,448 km, making up 22% of Brazil's territory (Ministry of the Environment Portal). In the state of Mato Grosso, the Cerrado region occupies 38.19% "covering mainly the Upper Paraguay-Guaporé depressions, the south and southeast of the Parecis plateau." (SCHWENK, 2005, p. 252). The author emphasises that the biogeographical formation of the Cerrado is classified according to its development: Clean Field; Dirty Field; Cerrado Field and Cerradão. Table 11 below shows the classification of the Brazilian Cerrado:

TABLE 11: Classification of the Cerrado.

NAME	IBGE CLASSIFICATION	FEATURES
Cerradão	Forested Savannah or Dense Savannah	Straighter trunks , and dense canopies.
Cerrado Field	Wooded Savannah or Open tree savannah	Constitutioncampestre , twisted trunks and branches.
Wall Cerrado	Savannah Park	Found in a variety of environments, from the wettest to the driest.
Field	Grassy Savannah	Found on the tops of
		plateaus flooding and wetlands.

SOURCE: IBGE (2012)
Organised by: SANTOS (2014).

The Cerrado has other nomenclatures according to the variation of authors, according to IBGE (2012), the main ones are Humboldt who presented the **term "Steppe"; Chevalier and Guénot used the name "Savannah" and the** Radam Brazil Project in the 1973-1987 survey with the term **"Savannah (Cerrado)".** Thus:

<blockquote>
The term Savannah (Sabana, in Spanish) is derived from the indigenous Caribbean term Habana (COLE, 1963, 1986; MARCHIORI, 2004) and, according to various authors, entered the phytogeographic literature through Fernández de Oviedo y Valdés (1851-1855), who used it to refer to the **"llanos" of the Orinoco Basin in northern South America.** [...] it was decided to adopt the term Savannah as a priority and Cerrado as a regional synonym, as it presents an ecological phytophysiognomy homologous to that of Africa and Asia. (IBGE, 2012, p. 108-109).
</blockquote>

With regard to the Pantanal, the vegetation is considered to be complex, i.e. it has a varied formation throughout the Pantanal region in four areas: those that are permanently flooded (the Pantanal plain); areas with flooded soils during the flood that don't dry out completely during the ebb; areas that are flooded periodically and areas of higher relief that don't flood (the BAP plateau). Thus:

Pantanal vegetation is therefore related to hydrotopographical factors and soil fertility, as well as anthropogenic action that directly influences the environment. According to Pott (1995), the main formations are: grassland, aquatic, savannah (woody grasses) and trees, which can be found just a few metres apart in the same area. However, the predominant vegetation in the Pantanal is grassland and savannah.

vary according to soil fertility. In non-flooded areas, Cerradão (larger, denser trees) prevails, as is the case in the Santo Onofre Settlement.

Despite the variety, "vegetation is more disciplined in sandy areas, where as a rule there are fields where there is flooding. In clayey areas, there tends to be more shrubs." (POTT, 1995, p. 06). Sandy clay soil is characterised as poor in deeper regions, but is fertile on the surface due to the decomposition of organic material from higher up.

The Santo Onofre settlement is located on what used to be a cattle ranch, which is why around 13 plots already had pasture, making it easier to raise cattle, because pasture formation requires spending on clearing and ploughing the soil, seeds, labour and time to grow the pasture, These factors make it difficult to follow a productive path on the rural property, as they are antagonistic to the economic condition of the settlers, as they are low-income families, which is why the establishments that didn't have formed pastures opted to cultivate the land.

The documentary research of the public deed of purchase and sale revealed that the ownership of the land for the purposes of Agrarian Reform took place through the purchase of the São Benedito farm, the owners the sellers **Quintino Marques and his wife Leozilza Rondom Marques and others, This "others"** (emphasis added) is because the farm had already been passed on as an inheritance to their three children. The purchase took place on 15 December 2005, and the purchasers were the settlers themselves, the intervening party in the purchase was the Land and Agrarian Reform Fund through the Land Credit financed by the Bank of Brazil. Figure 06 below shows the sequence of land negotiations for the establishment of the PA.

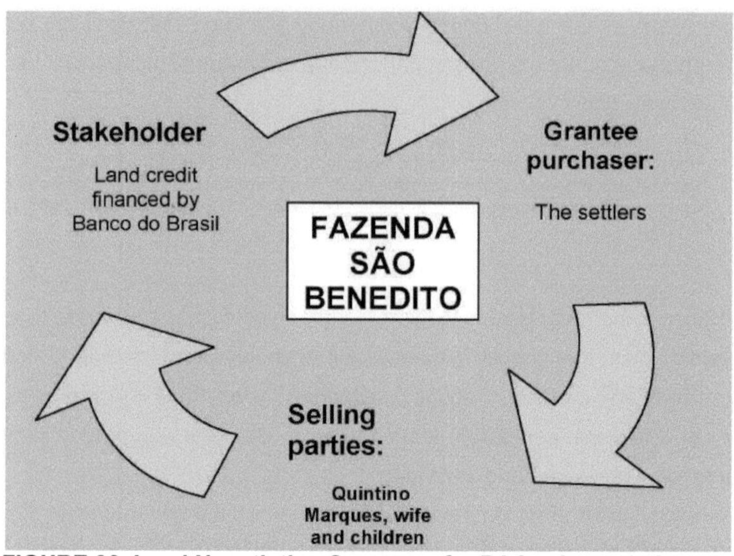

FIGURE 06: Land Negotiation Sequence for PA Implementation.
SOURCE: Public deed for the sale of property. Organised by: SANTOS (2014

The National Land Credit Programme (PNCF) is run by the Ministry of Agrarian Development (MDA) with the aim of providing financing for the purchase of land for those who don't have it or want to expand their rural establishment. According to information on the MDA website, in order to have access to land credit, as well as being a rural worker, you need to meet a number of requirements, namely: children of farmers or students at agro-technical schools; have five years' rural experience in the last fifteen years; have an annual income in line with that required by the line of financing.The PNCF's lines of financing are targeted according to the needs of the beneficiaries, of which there are three, as shown in table 12 below:

TABLE 12: PNCF Financing Lines.

LINE	DESTINATION	ANNUAL FAMILY INCOME	EQUITY VALUE	PAYMENT TERM	INTEREST RATE - YEAR
Combating Rural Poverty (CPR)	More rural families This can be used to purchase land and for community infrastructure projects.	R$ 9 thousand	R$ 15 thousand	36-month grace period. Up to 20 years to pay off the loan.	0,5%
Our First Land (NPT)	Aimed at young rural people, farmers' sons and daughters, students at agro-technical schools and family centres for work-linked training, aged between 18 and 29, who want to make their own life project viable in rural areas.	R$15 thousand	R$ 30 thousand	36-month grace period. Up to 20 years to pay off the loan.	1,0%

| Consolidation of Family Farming (CAF) | It assists farmers who are generally already on the land or those who have smallholdings and want to increase their area. The funds can be used to purchase land (SAT) and for basic investments (SIB), aimed at land, such as sharecroppers and tenants, for structuring production. | R$ 15 thousand | R$ 30 thousand | 36-month grace period. Up to 20 years to pay off the loan. | 2,0 % |

SOURCE: MDA Portal, Land Credit. Organised by SANTOS (2014).

The lot was purchased for R$28,540.00 and a total of R$40,000 was earmarked to finance the payment of investments and expenses beyond the purchase of the property, such as: improvements to the property; investments in basic structures; direct transaction costs; surveying expenses; environmental licences (when necessary); technical assistance expenses. Payment must be made to the PNCF partner, in the case of Santo Onofre the partnership was made through Banco do Brasil.

The deadline for repaying the loan for the Santo Onofre settlers was 16 years and 10 months, with a 22-month grace period followed by 15 annual instalments of R$2,666.67 plus interest. The first instalment was due on 10 September 2008 and the last instalment was due on 10 September 2022.

The settlement in question is part of the Consolidation of Family Farming (CAF) line of financing, in which case the grace period has already ended, but according to local leaders, only four settlers are paying, and the reason for the default is, according to the interviewees, the lack of financial conditions, This is because the income earned from the land is not enough to support the family and pay for the plot. In addition, the settlers mentioned that families from the agrarian reform programme are not allowed to leave their properties, which leads to a certain complacency on the part of local residents.

Defaulting on the payment of the plot directly affects obtaining resources and access to public incentive policies aimed at the settler, resulting in difficulties in remaining on the land, according to information obtained from the deed of sale of the property concerning defaulting borrowers:

> In the event of irregular use of the credit for speculative purposes, abandonment of the financed property, cessation of exploitation of the property or its disposal without prior and express authorisation from the UTE of the State of Mato Grosso, as well as any other irregularities considered to be intentional or unjustifiable or non-compliance with any other obligation arising from this contract, in addition to causing early maturity of this contract, the Borrowers [....] will be declared in default and unable to participate in any other MDA programme. (CONTRATO DE COMPRA E VENDA DE IMÓVEL- FINANCIAMENTO- PACTO ADJETO DE HIPOTECA, 2005, p. 04).

In 2013, Banco do Brasil called for the settlers to negotiate their debt on the plot, but the number of residents interested in paying did not exceed the four already mentioned. According to information from the local leader, SEDRAF (Secretariat for Rural Development and Family Farming) through the relevant sector of the UTE (State Technical Unit) proposed at a meeting of the settlement's residents' association that land ownership be regularised in 2014, both for the holders (the first settlers) and for those who bought plots, in order to provide payment for the land and legal ownership of it, as well as providing access to land credits.

The difficulties faced by small producers in accessing rural credit are related to the development of the institutional bases of public policies aimed at the countryside. The National Rural Credit System, created in 1965, was aimed at subsidising medium and large producers to invest in both the costing of production and marketing:

> Official rural credit, the main instrument used to promote the modernisation of agriculture, was highly selective, as its offer was restricted to medium and large producers. The vast majority of farmers, especially small landowners, tenants, partners and sharecroppers, whose conditions of access to land were precarious, were not assisted by official rural credit and found it more difficult to change the technical basis of production and remain in the countryside (HESPANHOL, 2007, p. 274).

Issues related to Land Credit as a means of agrarian reform raise questions about its applicability. The fact that it settles families does not mean their economic emancipation in rural areas. The fact is that in addition to the family group's expenses and production on the property, there are also costs related to land acquisition. The success of the land credit programme requires assistance from public policies that better structure rural workers, providing support and projects that increase family income so that they can survive and regularise their rural property.

The credit line made available to the settlers was Pronaf (National Programme for Strengthening Family Agriculture), which was used to build 26 dams, 2 irrigation systems, 19 simple wells and 5 semi-artesian wells. The issue of building dams and wells is necessary in the settlement due to its location in the upper part of the Pantanal, so during the dry season it is necessary to use dammed water to irrigate production.

Pronaf is intended to finance individual and collective projects to generate income for agrarian reform settlers and family farmers, and can be used to fund production or to purchase machinery and infrastructure on the property (Portal MDA). In order to access the credit, you need to have a regularised CPF and no debts, as well as obtaining a DAP (Declaration of Aptitude to Pronaf). There are seven lines of credit available, shown in table

13 below:

TABLE 13: Pronaf credit lines.

LINE	OBJECTIVE
Pronaf Funding	To finance agricultural activities and the processing or industrialisation and marketing of own or third-party production covered by PRONAF.
Pronaf Mais Alimento - investment	To finance the implementation, expansion or modernisation of production infrastructure and services, whether agricultural or non-agricultural, on the rural establishment or in nearby rural community areas.
Pronaf Agroindustry	Financing investments, including infrastructure, aimed at the processing and commercialisation of agricultural and non-agricultural production, forestry and extractive products, craft products and the exploitation of rural tourism.
Pronaf Agroecology	To finance investments in agroecological or organic production systems, including the costs of setting up and maintaining the enterprise.
Pronaf Eco	Financing investments in techniques that minimise the impact of rural activity on the environment, as well as enabling farmers to live better with the biome in which their property is located.
Pronaf Forest	Financing investments in projects for agroforestry systems; ecologically sustainable extractive exploitation, forest management plans, restoration and maintenance of permanent preservation areas and legal reserves and recovery of degraded areas.
Semi-arid Pronaf	To finance investments in projects to coexist with the semi-arid region, focused on the sustainability of agro-ecosystems, prioritising water infrastructure and the implementation, expansion, recovery or modernisation of other infrastructures.
Pronaf Women	To finance investments in credit proposals for women farmers.
Youth Pronaf	Financing investments from young people's credit proposals farmers.
Pronaf Funding and Commercialisation of Family Agroindustries	To finance the cost of processing and industrialising their own production and/or that of third parties, for farmers and their cooperatives or associations.
Pronaf Share	To finance investments for the payment of quotas by family farmers affiliated to production cooperatives or for use in working capital, costing or investment.
Rural Microcredit	Financing agricultural and non-agricultural activities of lower-income farmers, with loans covering any demand that can generate income for the family being assisted.

SOURCE: Ministry of Agrarian Development Portal. National Programme for Strengthening Family Farming (Pronaf). Organised by: SANTOS (2014).

According to the titular settlers interviewed, the value of the Pronaf was approximately R$17,000, but they were unable to explain how the Pronaf was paid out, showing a certain lack of interest in the subject, and there was also a lack of information about the applicability of the funding. This indicates a certain lack of structure in the screening process to select the families to be settled, which is later reflected in the abandonment of the rural environment by those who did not fit in, thus showing a utopian agrarian reform or one that lacks consolidation.

Of those interviewed, around nine bought the plot and have no access to Pronaf, because the land parcels are not being paid for and the default prevents them from withdrawing the DAP (Declaration of Aptitude to Pronaf), without which there is no access to funding, as explained above.

Although the purchase of plots from the agrarian reform is not permitted, in the case of the residents who bought the right to possession of the land, they paid the amount attributed to the improvements on the property, showing an interest in regularising the property and awaiting the appropriate measures from SEDRAF.

According to reports from the settlers themselves and local leaders, the residents who opted to sell their plots, when they received Pronaf, invested in applications that did not correspond to production and basic structures on the property, and this led to problems later on, as they were unable to wait for production and commercialisation, so they sold their properties and left the countryside. However, no settlers who sold their plots were interviewed to confirm the information described above, but we did observe the presence of plots with initial characteristics, without infrastructure such as fences, wells or house extensions.

The relationships that exist in the study area show results that lead us to reflect on the effectiveness of rural settlement projects. In the case of Santo Onofre, of the 95 settled families, 53 do not live there, 16 plots are completely abandoned, and of the remaining 37, only 8 have caretakers who live on the property in order to work for pay or even to look after the property to prevent its degradation. Figure 07 below shows a graph of the reality observed in the field in general as regards the occupation of the Santo Onofre Settlement.

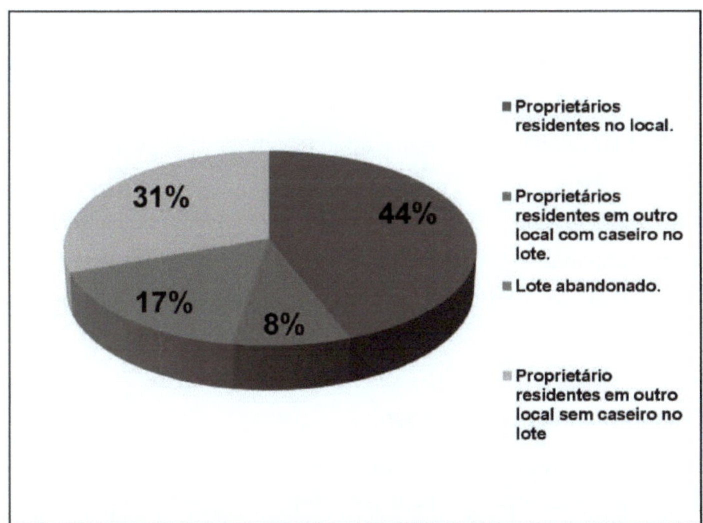

FIGURE 07: Occupation graph of the Santo Onofre settlement.
SOURCE: Settlement sketch. Organised by: SANTOS (2014).

According to the residents interviewed, the number of residents in the settlement is even smaller, but there are those who don't assume they live in the city. This information was confirmed in the field when we noticed that in some cases the houses we visited had no furniture, but the settlers denied this.

The case of living elsewhere and claiming to be a resident of the settlement is related to the fear of losing the right to possession of the land, because the information they have about agrarian reform refers to the removal of the property that is not developing rural labour and being passed on to another family on the agrarian reform waiting list.

The desire to live in the countryside is not enough, the family's livelihood prevails, so the alternative is to work elsewhere
returning to the property at weekends, with no time to devote to cultivating the land.

This raises questions about the effectiveness of Brazil's agrarian reform, which provides land but the settler has to pay for it, a process that provides land credits to be used for production and structuring the unit without providing the settlers with the necessary support and monitoring to ensure that the project is completed efficiently, procedures that reflect a land policy that "[...] is mistaken [...] and pushes the settlers into a sterile debate, conducted on an unequal basis." (PAULINO and ALMEIDA, 2010, p. 98).

6.2- Social Relations, Living Conditions and Production in the Santo Onofre Settlement.

There is a Residents' Association in the area, which until 2013 was inactive, but in

July of that year a new board of directors was sworn in to continue the association's work. As for the participation of residents, of those interviewed, 28 are members and those who don't participate, the reason given is the lack of interest on the part of the union in demanding benefits for the settlement, but according to the local leadership, the union's role is fulfilled, given the possibilities.

With regard to the social factor, the number of members per family is considered low. Analysing the data, it can be confirmed that some of the Santo Onofre families live in cities. Table 02 below shows the number of inhabitants of the families interviewed living in the settlement:

TABLE 02: Number of family members interviewed:

Number of homes	Number of components per home
3	1
8	2
8	3
7	4
2	5
3	Over 6
TOTAL: 31	

SOURCE: Interviews with residents of the settlement. Organised by: SANTOS (2014).

This can be explained by the lack of basic infrastructure in the area, such as a school building with primary and secondary education, resulting in students leaving for other places, difficulties related to health care, as there is no health care unit in the settlement for consultations, dependence on increasing income through work outside the area, lack of access to virtual means of communication such as the internet and telephone, as well as domestic comfort, as the structure of the houses is simple and unfinished.

Structures such as roads and electricity are in place, but these are aspects that do not guarantee the satisfaction of the residents. The EJA (Youth and Adult Education) and primary school classrooms are in the house destined for the settlement's headquarters (Figs. 08, 09, 10 and 11). The precariousness is visible, which does not make it an attractive environment for those who want to study; the following figure shows the current situation of the settlement's headquarters and classrooms:

Figure 08: Headquarters house used as a school.
Figure 09: Details of the broken window in the head office.

Figure 10 and 11: The precarious structure and disorder of the classroom on the premises.

PHOTOS: SANTOS (2014).

Analysing in general, there is a certain lack of interest on the part of the government in investing in improvements in the countryside, especially in the area of education, because peasant subordination and the domination of the capitalist classes have existed since the beginning of the formation of the territory of Brazil, the system in force tends to grow at the expense of the lower classes, and this exploitation is seen in reality

Regarding the structure of the houses built in the settlement, they have four pieces (Figs 12 and 13) and a bathroom, without any finishing, the other buildings on the plots were built with PRONAF funds. Below you can see the different building structures on the plots in the area surveyed:

Figure 12: Property without infrastructure, house showing characteristics of early construction without finishing.
Figure 13: Property with infrastructure such as fence, orchard, finished house.
PHOTOS: SANTOS (2014).

Another problem is the totally abandoned plots (Fig. 14). For residents, this scenario reflects the lack of supervision by the competent bodies. A system should be set up to check the occupation of plots and, if it is found to be abandoned, immediately arrange for new settlers to occupy them, thus guaranteeing land for those on the agrarian reform waiting list.

Figure 14: Abandoned plot with vegetation all over the property. PHOTO: SANTOS (2014).

The tools used to work the land are hand tools such as hoes, crawlers, machetes, among others, and most of the handles are handmade by the settlers themselves, taking wood from the local forest, as well as other objects of domestic utility, so it can be seen that the domestic industry is part of the reality of the local people.

To better understand the term domestic industry, we point out what Ribeiro et al (2012, p. 04) mention about the concept, citing that "[...] it comes from classic authors of economic literature. Fernandes and Campos (2003) indicate that the concept was used by Marx, Kautsky and Lenin to designate the manufacturing activities carried out in peasant production units [...]".

These handmade objects are part of the culture inherited from father to son; producing tools is a way of saving money for the peasant, as it avoids parallel spending. Figures 15, 16, 17 and 18 show examples of utensils made by the Santo Onofre settlers:

Figure 15: Pestle and wooden spoons.
Figure 16: Benches made from local wood.
PHOTOS: SANTOS (2014).

Figure 17: Bowl, trough and table.
Figure 18: Cassava press for making flour.
PHOTOS: SANTOS (2014).

With regard to the age group of the settlers, it can be seen that the number of residents over 56 years old dominates (Fig. 19), a fact that highlights the reality of the return to the countryside, since older people have already had a relationship with the countryside in their childhood or youth, living in the city for some time they have not forgotten rural life, the main reason for their return to rural establishments

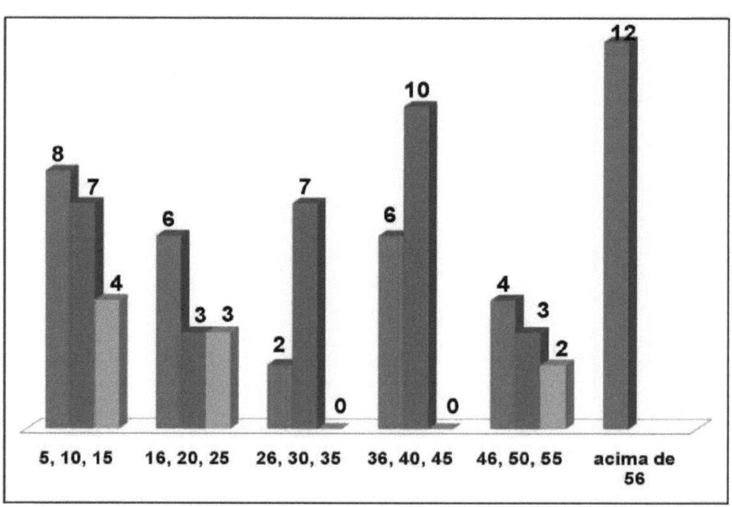

FIGURE 19: Graph of the Approximate Age Range of Santo Onofre Residents.
SOURCE: Interviews with settlers.
Organised by: SANTOS (2014).

As Paulino and Almeida (2010) point out, this is the process of recreating the peasantry, which came to prominence through conflicts in the 20th century in favour of the right to own land, such as in Mexico in 1910, Russia in 1905 and 1917, Cuba in 1958, and so on.

phenomena requires dedication on the part of researchers, academic production raises analyses that show the old and recent lessons about the peasantry. So:

> [...] the recamponisation of the landless, most of whom are city dwellers, raises new questions, especially those linked to the task of knowledge, of peasant knowledge as a possibility of overcoming crises. Thus, the fundamental thing for those who **are** recreating themselves in contemporary times as peasants is not to **be** peasants, but the character of their social existence (PAULINO and ALMEIDA, 2010, p. 57).

The phenomenon of people returning to rural areas, especially retired people, has been adding to the number of people living in the countryside, especially in places close to medium-sized and large cities. This return migration must be analysed when formulating policies aimed at increasing projects in rural areas (HESPANHOL, 2007).

Income from pensions or benefits guarantees the family's food, raising the question of income generation on the property, leading us to understand that there are contradictions in the planning of agrarian reform in Brazil. Projects aimed at rural areas are supposed to provide rural dwellers with income from rural property, but in reality, income from other areas is seen as a guarantee of the family group's livelihood.

In order to improve the performance of activities in the countryside, it is necessary to make use of the labour force of the young and adult classes, making activities more dynamic and thus obtaining results that reflect an increase in the family group's income. That's why we need to take a careful look at the consolidation of agrarian reform projects. Therefore:

> Maintaining the rural population is a challenge that must be tackled by establishing public policies capable of generating income and thus guaranteeing the social reproduction of this category of producers (HESPANHOL, 2007, p. 278).

The level of education of those interviewed is considered low: 16 residents have only incomplete primary education, only two have a university degree and around ten interviewees are studying at secondary school.

EJA modality in the settlement itself this year and only three declared themselves illiterate.

As for the origin of the residents, it was found that most of them are from Mato Grosso, mainly from the municipality of Poconé, which accounted for 13 of the interviewees, while only seven reported being from other states, namely: two from Paraná and São Paulo; one each from Ceará, Minas Gerais and Mato Grosso do Sul.

The presence of Poconeans in the settlement leads us to reflect on the land ownership structure in Mato Grosso's Pantanal, which presents a picture of land concentration inherited from the sesmarias period in Brazil:

> The seasonal climate and the terrain conditions that caused part of the properties to be flooded at certain times of the year influenced the size of the properties and legitimised the large size of the farms in the Pantanal, which often had no dividing fences between the establishments. Owners rarely had title deeds as a material document, but their boundaries were respected by their neighbours and marked by geographical features (ROSSETTO E GIRARDI, 2012, p. 142).

The processes that followed the sesmarias, such as the legitimisation of ownership and regularisation of land, when land became private property, remained the concentration of large establishments. This situation is still a reality in the Pantanal of Poconé, and despite the development of settlement projects, there are still idle people who want access and the right to cultivate on their own land.

6.3- Pluriactive Activities as a Way of Obtaining Income in the Santo Onófre Settlement.

The rural environment is increasingly attracting attention with regard to the activities carried out in it, the relationship with the land and, above all, staying in the countryside.

In order to better understand the struggle for survival in the rural environment, it is necessary to know the forms of activity carried out by those involved in this reality. In this way, we can analyse the ways and means used, in this case by the small rural producers of the Santo Onófre Settlement, to obtain income and remain in the settlement.

Pluriactivity is present in the experience of the settlers in question, so an approach to this subject is necessary in order to cover the research topic. For their economic sustainability, the practice of pluriactive activity to increase the income of the settlers is often necessary, because for most of the residents of the settlement, the income obtained from the land alone is not enough to maintain the family's expenses, so accessory labour is indispensable. According to Dantas et al (2012, p. 03):

> In Brazil, discussions on pluriactivity in family farming began systematically in the mid-1990s with studies by Schneider and Sacco dos Anjos in 1994 and 1995 in the southern region, which portrayed typical situations of part-time farming practised by family farmers in southern Brazil.

This situation, according to Sacco apud Dantas et al (2012, p. 04), has become "[...] a

significant emergence and/or expansion of non-agricultural activities in the national countryside and the pluriactivity practised by rural residents." This is due to the need to survive in the countryside.

The reality expressed in the countryside, in which the income obtained from the land is not compatible with the expenses of the small producer's family, pluriactivity is a viable alternative to supplement this income. In this way, we can say that the concept of pluriactivity is a kind of meeting of agricultural activities with other activities that provide income or not, involuntarily, whether internal or external to agricultural production:

> A social reproduction strategy used by agricultural units that operate fundamentally on the basis of family labour, in contexts where their integration into the social division of labour does not stem exclusively from the results of agricultural production, but above all through the use of non-agricultural activities and through links with the labour market. In this sense, the author argues that, although integrated into the social and economic order, these family units find spaces and mechanisms not only to survive, but to assert themselves as a social form of organising work and production with multifaceted characteristics. (SCHNEIDER apud DANTAS et al, 2012, p. 05).

Therefore, pluriactivity is a means of economic support for small rural producers. In order to better understand the subject in question, it is important to understand accessory labour.

When the family has more labour than the activities carried out on the farm, or when the produce acquired from the farm doesn't meet the needs of the family group, the family members look for another form of income, a salary, by working in agriculture or in other activities, on or off the farm, i.e. accessory labour, as confirmed by Chayanov apud Girardi (2008, p. 98):

> When land is insufficient and becomes a minimum factor, the volume of agricultural activity for all the members of the farm is reduced proportionally, to a varying degree but inexorably. But the labour force of the family operating the unit, when it can't find a job on the farm, turns [...] to craft, commercial and other *non-agricultural activities in* order to achieve an economic balance with the family's needs.

This scenario of the search for economic sustenance on small rural properties is increasingly evident today, given the capitalist mode of production that governs the country's economy, driving the expropriation of many small properties which, because they don't have the means to enter this system, tend to look for other forms of income to provide for the family, working outside the productive sphere of the property.

On the subject in question, the Santo Onofre settlers practice accessory labour, since, according to them, the income from the property is not enough to support the family. The main types of work carried out by the interviewees can be seen in table 14 below:

TABLE 14: Economic Activities Developed by Residents Interviewed in the Settlement.

TYPES OF WORK	NUMBER OF INTERVIEWEES	WORKPLACE
Manual labour	5	In the settlement and properties neighbours.
Mining	2	Gold mining near the settlement.
Construction area (bricklayer)	4	Nearby towns (Poconé, Livramento, Várzea Grande and Cuiabá); on farms and on site.
Professor	2	In the settlement and in the District of Cangas - Poconé.
Various activities	6	Poconé and Cuiabá.

SOURCE: Interviews with settlers.
Organised by SANTOS (2014).

When asked about off-farm activities, there was confirmation from the literature. The answers unanimously referred to the issue of insufficient income on the property, the lack of help from public policies that support both the cultivation and commercialisation of production, economic insecurity and the lack of conditions to wait for the land to produce are among the main concerns cited by the settlers interviewed. This leads us to understand that, in fact, without political support, agrarian reform cannot be consolidated. It's not enough just to provide the land, but also to promote the conditions for remaining on it.

6.   4-Production relations   and the Santo Onofre settlement.

With regard to the peasant economy of the properties in the settlement, it was noted that the main crop grown is pineapple, a monocotyledonous plant of the bromeliad family, subfamily Bromelioideae. The pineapple trees belong to the *Ananas comosus* species, which includes many fruit varieties. In the case of Santo Onofre, it's the pearl pineapple, present in 16 properties, destined for commercialisation, considered the "flagship" (emphasis added) in the extraction of income on the property, considered the most profitable and easy to negotiate.

There are other types of agricultural production in the area. Among the interviewees, 14 properties grow a variety of vegetables, 03 grow sugar cane, and 20 grow varieties such as manioc, okra, watermelon, pumpkin, maize, etc.

The Poconé municipal government provides a small tractor for ploughing the land on small properties, and only the fuel used by the tractor is counted as an expense for the farmers. The issue, according to the settlers, is the difficulty in obtaining this service, because the bureaucracy is huge and the settler is often not attended to when he needs it.In the context of production in the settlement, it was seen *on site* that the settlers grow crops mainly for family consumption, while surpluses are sold to supply other products that are not extracted from the land itself; among the main subsistence crops is manioc. Figures 20 and 21, respectively, show the graphs with the results obtained from the interviews regarding agricultural products for subsistence and those for sale:

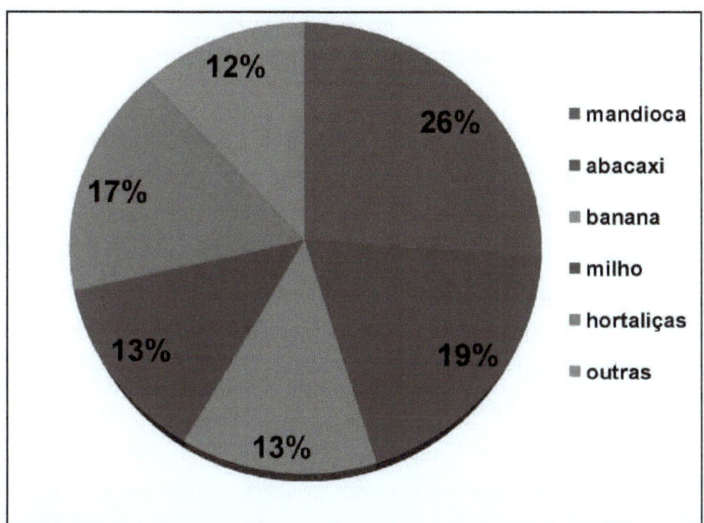

FIGURE 20:   Graph of Subsistence Agricultural Products in the Local Rural Establishments.
SOURCE: Interviews with settlers.
Organised by: SANTOS (2014).

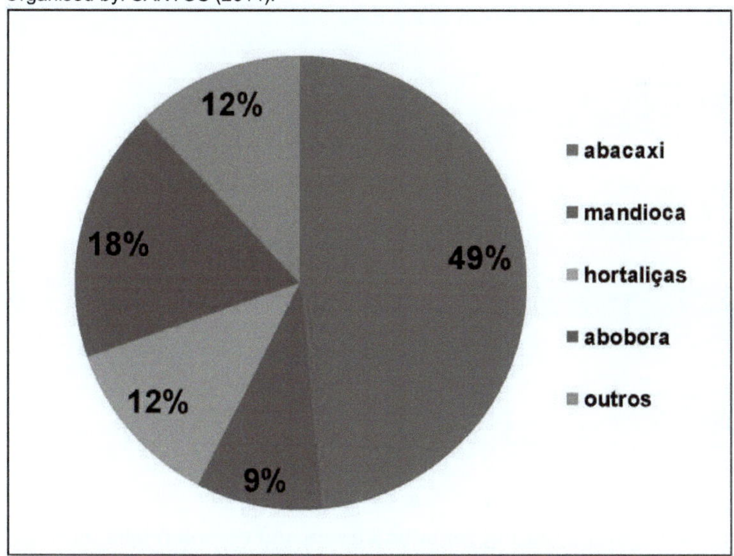

FIGURA 21: Graph of Agricultural Produce for Sale.
SOURCE: Interviews with settlers. Organised by: SANTOS (2014).

The complaints made by the residents of the Santo Onofre Settlement regarding production are linked to the difficulty in obtaining a price when selling their products. The devaluation by traders when negotiating discourages producers, leading them to reflect on whether it is worth producing to sell for such a low price. Another complaint concerns the lack of a co-operative in the settlement, which would make it easier for the settlement's producers to have an appropriate place to sell their produce.

The question then arises again of the applicability of resources in rural areas by the government, such as setting up co-operatives, enabling small producers to trade their produce. In the case of Santo Onofre, where agricultural activities are practised, it is necessary to invest in the destination of production, thus offering an incentive for the settlers to continue in the countryside, earning income from the establishment.

With regard to animal husbandry on the site, small-scale livestock farming takes place in 12 establishments (totalling 148 head) and 29 with poultry and pigs. The main purpose is for the family's own consumption, and when there is a surplus, it is sold in the settlement. As well as cattle, other animals are reared on the farms, namely: chickens, pigs, sheep and fish, but only nine farms use their production for commercialisation, in this case, one fish farm, two pig farms, five poultry farms and one cattle farm. The number destined for commercialisation is much lower than for subsistence, thus confirming the readings made prior to the field research about the peasantry, whose characteristic is production for subsistence. Figure 22 below shows the graph that proves this:

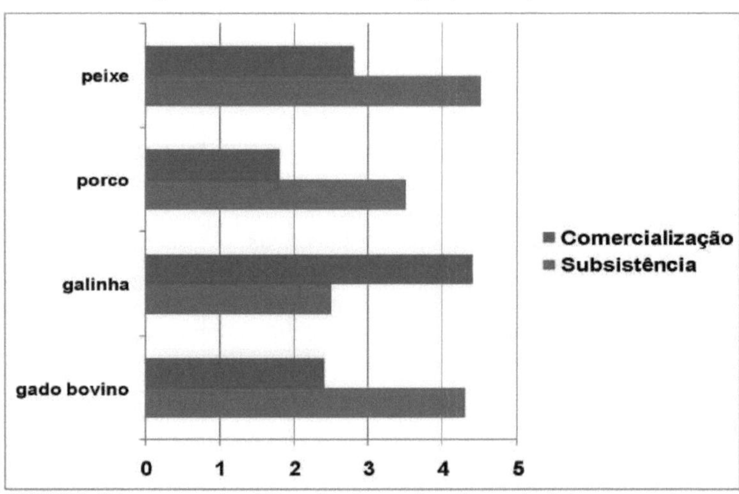

FIGURA 22: **graph of livestock rearing for subsistence and commercialisation on farms.**
SOURCE: Interviews with settlers.
Organised by: SANTOS (2014).

As can be seen from the graphs presented, it can be seen that the main way of earning income on the property is not by managing animals, but by farming, mainly pineapples. The use of milk extracted from cattle for marketing purposes is also considered low, as there are only four establishments that sell milk and another three make cheeses (Fig. 23) and sweets to sell in the settlement itself.

FIGURE 23: Cheese made by a local resident. PHOTO: SANTOS (2014).

There is also a domestic flour mill in the settlement, where the cassava flour produced basically uses family labour. It was formed through the partnership of two neighbouring landowners, who also share their livestock using pasture from the two establishments. The income from the sale of flour is shared in half between the partners, and in the case of cassava supplied by third parties, the profit is also shared equally between the manufacturer and the supplier of the raw material. The profit is considered reasonable to maintain the families, but there is still a need to work outside the establishment with manual labour to increase income. Figures 24 and 25 below show the manioc grinder and oven and the handmade utensils used to make manioc flour.

Figure 24: Cassava grinder used in the domestic flour mill.
Figure 25: Oven and handmade utensils for toasting manioc flour.
PHOTOS: SANTOS (2014).

The income obtained varies according to the harvest period. During the pineapple season, it is satisfactory, as the number of pineapple trees varies between 10,000, 20,000 and 30,000, with only one property having 140,000 pineapple trees. The profit is generally invested in the establishment itself and in buying household utensils, but for the rest of the year, the family basically survives on a minimum wage.

It can thus be seen that the logic of peasant production exists in the links found at the research site, where the merchandise produced has no other purpose than to supply the property and the family group, confirming the Marxist theoretical bases on the peasant economy in the form of M-D-M (merchandise-money-merchandise).

Generally speaking, the origins of local landowners' forms of income vary: there are those who work solely on the property, receive some kind of benefit as well as having homes in other localities, others are caretakers in local establishments, table 15 below shows the economic aspects of the families interviewed in the settlement:

TABLE 15: Economic Aspects of the Families Interviewed in the Santo Onofre Settlement.

Working relationship	Employees - 25% Self-employed - 50% Work on the property - 25%
Monthly income	1 to 2 salaries - 46% 2 to 3 salaries - 36%

	Above 3 salaries - 18%
Benefits	19% of the residents receive benefits, of which 53% are pensioners, the rest are divided between pensions, family grants and school grants.
Homemakers' income	Of the five caretakers interviewed, two earn the minimum wage, two earn less than the minimum wage and one works on a daily basis (40 reais a day).
Townhouse	39 per cent reported having a house in the city, mainly in Poconé and Cuiabá, which they rent out or use as a family home.

SOURCE: Interviews with settlers. Organised by: SANTOS (2014). .

Living in the countryside for the residents of Santo Onofre is considered arduous, but all the reports expressed the desire to remain in rural areas and the clamour for better living conditions, support from public policies and adequate infrastructure were unanimous among the interviewees who actually live in the settlement, and were pointed out as necessary to remain on the property.

6.5- Documentary Research on Projects Presented to Settlers.

Documentary research carried out on minutes made available by a local leader made it possible to identify projects presented to benefit local residents. In short, at least five proposals were observed which, if consolidated, could add income to residents.

At first, in 2009, with the start of the Peasant Women's Association in the settlement (deactivated since 2012 due to lack of members), at a meeting with the secretary of agriculture of the municipality of Poconé at the settlement's headquarters, a project was presented to work with fruit pulp, given that pineapple is grown to a greater extent in the settlement. For the production of juices, the secretary of agriculture was promised a pulper (a machine that removes the pulp from the fruit and separates it from the peel or stone), which is necessary for the production of juices on a larger scale.

Among the complaints made by the settlers is the issue of unfulfilled promises. To date, no proposed project has come to fruition, causing a certain amount of disbelief among the small rural producers who are settlers under agrarian reform as to the conclusion of projects to support their work in the fields. The survey found that the following projects had been offered to settlers, see table 16 below:

TABLE 16: Projects presented to the Santo Onofre settlers.

Project	Objective	Conclusion
Pulper; Sealer and Electronic Scale	Produce juices on a large scale; Sealing pulp and juice containers; Weighing products accurately.	Abandoned
Tractor with plough and trailer, also a planter.	Make work easier and increase the cultivation	Abandoned

	area.	
Vacuum cassava	Destination of the manioc grown in the settlement for the vacuum manioc manufacturing industry.	Abandoned
Fish	Developing fish farming for small rural producers.	Abandoned
Farinheira	Making cassava flour in the settlement to add to the family income	Not finalised

SOURCE: Minutes of the Santo Onofre Residents' Association. Organised by: SANTOS (2014).

The projects mentioned above would, if realised, serve to structure the income of the settlers, but so far they have not materialised. There is widespread discontent among the families living in Santo Onofre, and despite numerous meetings to discuss the proposals, nothing has materialised so far.

In the case of the fish farming project, the land would first have to be analysed in order to open dams, with the understanding that only those who did not already have dams on their plots would receive support. The licences and bases for starting fish farming were the subject of discussions after the project was presented, and since there was no investment, the other projects presented fell into disrepute among the settlers.

In the case of the farinheira, this is a project funded by FINEP (Fund for Studies and Projects) aimed at the cassava production chain, with the necessary structure for the manufacture of cassava flour being installed in the selected locations. Despite the importance of the project, it was realised that many settlers showed disbelief, as the project has not yet been completed.At a meeting with those responsible for the project, the producers questioned what they would do with the large quantities of cassava if the flour mill wasn't installed. However, encouragement from the president of the settlement's residents' association, together with a representative from SEDRAF (the Secretariat for Rural Development and Family Farming), stimulated the acceptance of the settlers, who were willing to plant cassava and wait for the project to materialise.

6.6- Pineapple Production Chain.

Knowledge of the techniques used to grow pearl pineapple came first from EMPAER, the organisation responsible for technical assistance in the area. However, for residents who acquired plots later, the instructions are passed on by the settlers themselves. Figure 26 of the flowchart below shows the sequence of pineapple production up to the consumer.

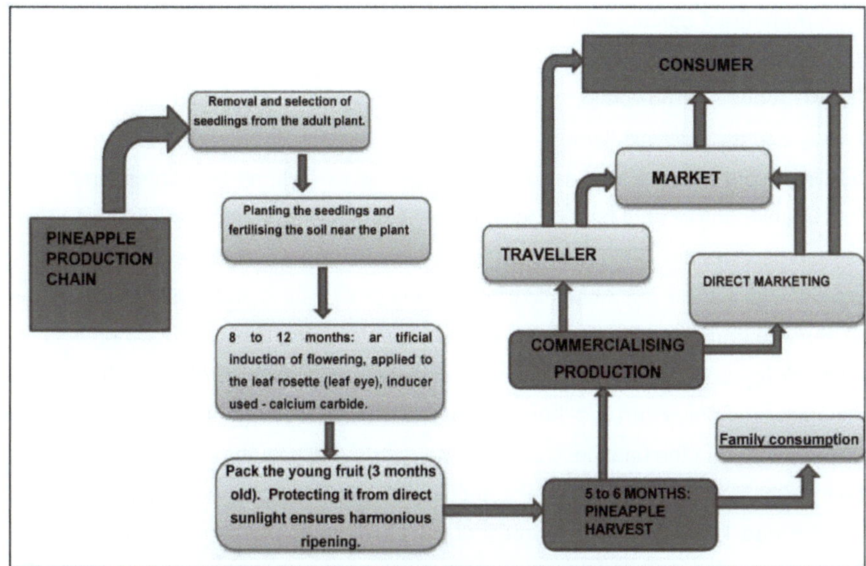

FIGURE 26: Flowchart of the Pineapple Production Chain.

SOURCE: Pineapple producers from the Santo Onofre settlement. Organised by: SANTOS (2014).

Pineapple seedlings are taken from the adult plant after the fruit has been harvested. There are three types of seedling: those born close to the ground at the foot of the plant, another below the fruit and the crown (located at the top of the pineapple). However, in local cultivation, seedlings born below the fruit are used and go through a rigorous selection process, followed by planting those that are of good quality, coming from healthy adult plants to guarantee better fruit.

Also, after planting, fertiliser is deposited in the soil next to the plant, strengthening the growth as well as the development of the fruit. O

he pineapple usually starts to produce naturally after one year and six months, but this process is advanced by producers with the use of inducers. At this point, the pineapple tree receives a small amount of calcium carbide which is added to the eye of the plant. This method speeds up the fruit setting period. The carbide season takes place when the plant is between eight and nine months old, giving the producer half a year's advance on the harvest.

Another important aspect is the time it takes to "wrap" the pineapple when it is halfway through its growth and ripening process, which happens after three months of fruiting. The term "wrapping" is used by the settlers to mean the act of wrapping the fruit in newspaper to protect it from direct contact with the sun's rays, providing harmonious ripening on both sides for better commercialisation.

After six months of flowering, the fruit is ready to be harvested, and it is sold in two ways: directly and through middlemen.

The direct sale of produce is done by the producer himself, he negotiates the market without the intervention of third parties, in which case the owner has the means of transport to sell the harvest. The presence of middlemen in the settlement is considered important by local producers, as without the means to transport the produce it is profitable to sell it on the plot itself, thus guaranteeing commercialisation and avoiding the loss of production.

Fieldwork revealed the presence of at least three middlemen, all of whom have plots in the settlement, two of whom live there and one of whom does not. Interviews with them provided information on the final destination of the goods and how they were traded.

The middleman, who buys in large quantities; despite having an establishment on the site with more than 140,000 pineapple trees, doesn't live on the plot. The products are sold at fairs in Cuiabá, as well as to traders at the Porto market, according to reports from the settlers.

In another case, the product traded is destined for the cooperative in Poconé, because the buyer previously lived in a traditional community and had access to deliver goods directly to the cooperative.

Finally, the products bought in the settlement are sent by another middleman to markets in Cangas, Poconé, Várzea Grande and Cuiabá. In the case of the last two, they both live in the settlement and buy and sell goods as a form of income.

It was found that pineapple production is considered more profitable in Santo Onofre. The fruit is usually sold in abundance, which guarantees a greater amount of money over the same period, pineapples require less work to grow, expenses are also low, which makes the producer enthusiastic about the crop. It should be noted that small producers tend to produce crops that do not require labour expenses, apart from those related to cultivation, as this way the profit becomes greater, favouring the farmer.

CHAPTER 7

FINAL CONSIDERATIONS

Studies on peasant socioeconomics are important for understanding the links that exist in rural areas. When approaching social and economic relations in smallholdings, one can see similarities in terms of their development and survival in the countryside.

The Brazilian agrarian question reflects a past of exploitation by the Portuguese colonisers, the accumulation of land in the formation of large estates from the beginning to the present day is still a reality. Throughout Brazil's agrarian history, large estates have been strengthened by the capitalist economic model in force in most countries, including ours. On the other hand, the struggle to gain access to land is evident, and rural social movements have played an important role in this process.

The development of agrarian reform settlements triggers a series of analyses as to the authentic condition of the Brazilian peasant. The search for a decent quality of life on the land is a reality for many of those who live in the countryside.

Government policies from the 1970s onwards promoted a slight acceleration in Brazil's land restructuring. The programmes inherent to Agrarian Reform, although slow, promote access to land for those who don't have it.

The support of public policies such as the availability of land credits also helps to strengthen small rural establishments, but they are not enough to guarantee their permanence in the countryside. There is a need for assistance to farmers in terms of basic structures such as schools; health centres; stable roads; access to cooperatives; the provision of subsidies, among others, which contribute to their continuity on their properties.

The results consolidate our understanding of Brazil's agrarian reality. Peasants have similar characteristics to other regions of Mato Grosso, especially in terms of survival and staying in the countryside. This implies that greater investment is needed on the part of public authorities to provide better conditions for those who produce on a small scale, harmonising their experience in rural areas.

As far as the local economy is concerned, the search for income outside the property is common among the settlers, receiving a salary is seen by the residents as a guarantee of food, the fact of working as an accessory provides economic stability for the family group, and the production from the property serves, in this case, to help with household expenses.

This leads us to understand that there are flaws in the so-called agrarian reform, because one of its aims is to provide a livelihood for the settlers acquired from the rural establishment, which in reality does not happen. The interviewees unanimously reported that it is difficult to derive family income from the land alone, and that it is essential to carry out other activities outside the property.

In the research area, it was observed that the subsistence production of various crops (polyculture) sustains the family's food security, minimising expenditure on groceries such as meat, vegetables, among others. The art of making useful tools also contributes to the residents' economy. Objects such as tool handles, tables, stools, spoons, pestles and bowls are commonplace in homes, which means that they don't have to spend money buying them.

Based on the results obtained in the field, it can be said that the economic situation of the residents of Santo Onofre is the result of hard work on the rural property, as well as the difficulty of entering and competing in the market. However, those who chose to stay in the settlement were persistent, so we can say that the settlers elevate their feelings for the land above the economic factor, because the relationships they experience overcome the obstacles imposed by the capitalist economic system in force in the country.

It's important to mention that we considered analysing the residents of Santo Onofre as peasants because, although they are present in the market, they have intrinsic characteristics of peasant logic, maintaining relations in the establishment as working land, and not as business land as the capitalist mode of production leads.

Despite the claim that peasants are tending to disappear, however, we consider that they are still fighting to gain access to land in many parts of Brazil, including the state of Mato Grosso and the Pantanal region of Poconé.

BIBLIOGRAPHICAL REFERENCE

ANDRADE, Maria Margarida de. **Introdução à Metodologia do Trabalho Científico.** São Paulo: Atlas, 1997. v. 2.

BARCELLOS, Sérgio Botton. **Agrarian Reform in Brazil in Question: a debate off the agenda?** Opinião Pública, 2013. Available at: <http://www.sul21 .com.br/jornal/2013/05/a-reforma-agraria-no-brasil-em-questao- um-debate-fora-das-pautas/> Accessed on: 08 August 2013.

BRAZIL, Law 11.326 of 24/07/2006. Establishes the guidelines for formulating the National Policy for Family Farming and Rural Family Enterprises. Available at: <http://www.planalto.gov. br/ccivil_03/_ato2004-2006/2006/lei/l 11326. htm >

. Law 4.504 of 30 November 1964. Provides for the Land Statute and other measures. Available at: <http://www.planalto.gov.br/ccivil_03/leis/L4504.htm. Accessed on: 27 June 2013.

. Law 601 of 18 September 1850. Provides for the Empire's vacant lands. Accessed on 14 May 2013. Available at: <http://www.planalto.gov.br/ccivil_03/Leis/L0601-

1850.htm> Accessed on: 30 June 2013.

. Law 4.214 of 02/03/1963. Provides for the "Rural Labour Statute. Available at: <http://www.planalto.gov.br/ccivil_03/leis/1950-1969/L4214.htm> Accessed on: 27 June 2013.

. Decree 91.766 of 10 October 1985. Approves the National Agrarian Reform Plan - PNRA, and makes other provisions. Available at: <http://www2.camara.leg.br/legin/fed/decret/1980-1987/decreto-91766-10-outubro- 1985- 441738-publicacaooriginal-1 -pe.html> Accessed on: 04 July 2013.

BARUFI, Clara; SANTOS, Edmilson dos; IDE, Cristiane. , **Energy self-sufficiency and development: the natural gas trade between Brazil and Bolivia**, year 5, vol. 2, Cadernos PROLAM/USP. São Paulo: 2006. Available at: <http://www.usp.br/prolam/downloads/2006_2_6.pdf> Accessed on: 15 March 2014.

BECKER, D. ; MILTON, L. **Regional Development: Interdisciplinary Approaches.** SANTA Cruz do Sul: EDUNISC, 2003.

CARVALHO, Kelly Cristina. **Cattle Production Chains of Medium and Large Properties in the Northern Pantanal of Mato Grosso, a Study in Cáceres-MT.** Dissertation (Master's) - Postgraduate/Master's Programme in Geography, ICHS, Federal University of Mato Grosso, Cuiabá, 2012.

CARVALHO, K. C; LIMA, D. M. D. F. de; ROSSETTO, O. C. **Labour in the countryside:** formality and informality on rural properties in Cáceres, Mato Grosso, Brazil. 5th Symposium on Natural and Socioeconomic Resources of the Pantanal - Simpan/2010.

BUARQUE, Sergio José Cavalcanti. **Building Sustainable Local Development: planning methodology.** Rio de Janeiro: ed. Garamond, 2002.

CHAYANOV, A. On the theory of non-capitalist economic systems [1924]. In: SILVA, J. G. da; STOLCKE, V. (Org.) **A Questão Agrária: Weber, Engels, Lenin, Kautsky, Chayanov, Stalin.** São Paulo: Brasiliense, 1981 - Available at : <http://www.revistacarbono.com/artigos/04agricultura-camponesa-paulopetersen/#sthash.nXl1eBFL.dpuf> Accessed: 20 May 2013.

DANTAS, Lucivalda Sousa Teixeira e. et al. Pluriactivity in Family Farming: weaving income and (re)constructing identity? In: Encontro Nacional de Geografia Agrária, n. 21, 2012, Minas Gerais. **Proceedings.** UFU: LAGEA, 2012. 1438 - 1. Available at: <http//www. lagea.ig.ufu.br/xx1enga/anais_enga_2012/eixos/1438_1 .pdf> Accessed on: 18 October 2013.

DELGADO, A. C.; LEMGRUBER, S.. **Indigenous Movements and their Implications for the Political Process in Bolivia and Peru.** Observador Online - South American Political Observatory, V. 1, n. 4, Jun. 2006.

DEVEZA, Felipe . **When the land is the limit - Zapata and the Mexican Revolution.** A Nova Democracia, Rio de Janeiro, 01 Dec. 2011.

FERNANDES, B. M.. The agrarian question: conflict and territorial development. In:

BUAINAIN, A. M. (Ed). **Land Struggles, Agrarian Reform and Conflict Management in Brazil.** Campinas: Editora da Unicamp, 2005.

FERNANDES, B. M. (org.) **Campesinato e Agronegócio na América Latina: a Questão Agrária Atual.** São Paulo: Expressão Popular. 2008.

FERRANTE, V. L. S. B. **The Rural Labour Statute and Funrural.** In: Resumos SBPC, Suplemento de Ciência e Cultura, 1975. p. 188-202.

FIGUEREDO, O. A. T.; MIGUEL, L. A. **Agriculture, environment and rural development: the 2nd Department of San Pedro, Paraguay.** Ribeirão Preto: 2005.

GIRARDI, Eduardo Paulon. **Theoretical-methodological proposal of a critical geographic cartography and its application in the development of the atlas of the Brazilian agrarian question.** Presidente Prudente, 2008.

GOMES, Mércio. **Evo Morales Initiates Agrarian Reform in Bolivia.** Rio de January: 2009. Available at: <http://merciogomes.blogspot.com.br/2009/03/evo-morales-inicia-reforma-agraria- na.html> Accessed on 15 March 2014.

HAESBAERT, R. **Region: paths and perspectives**. Paper presented at the First Conference on Comparative Regional Economics, FEE-RS, Porto Alegre, Department of Geography, Fluminense Federal University, 2005.

HESPANHOL, A. N.. The development of the countryside in Brazil. In: FERNANDES,B. M.; MARQUES,M. I. M.; SUZUKI, J. C. (Org.). **Geografia Agrária: teoria e poder.** 1 ed. São Paulo: Expressão Popular, 2007. p. 271-287.

HOFFMANN, R. **Income distribution: a measure of inequality and poverty.** University of São Paulo Press. 1998

INNOCENTINI, Thaís Cristina. **Hereditary Captaincies: Colonial Heritage on Inequality and Institutions.** Master's thesis - São Paulo School of Economics, São Paulo, 2009.

BRAZILIAN INSTITUTE OF GEOGRAPHY AND STATISTICS - IBGE. **Technical Manual of Brazilian Vegetation.** 2012. Available at: <ftp://geoftp.ibge.gov.br/documentos/recursos_naturais/manuais_tecnicos/manual_tecnico_vegetacao_brasileira.pdf> Accessed on: 05 January 2014.

NATIONAL INSTITUTE FOR COLONISATION AND AGRARIAN REFORM. **SIPRA report.** 2012. Available at: <http://www.incra.gov.br/index. php/reforma- agraria-2/projetos-e-programas-do-incra/relacao-de-projetos-de-reforma- agraria/file/1115-relacao-de-projetos-de-reforma-agraria> Accessed on: 03 June 2013.

KAUTSKY, K. **The agrarian question.** São Paulo: Nova Cultural, 1986 [1899].

LA BLACHE, V. de apud CARVALHO, G. L. Região: a evolução de uma categoria de análise da geografia. **Boletim Goiano de Geografia**, vol. 22, n° 01, jan./jun. 2002.

LAMERA, J. A.; FIGUEIREDO, A. M. R.; **Rural Settlements in Mato Grosso.** Cuiabá: UFMT, 2007.

LÊNIN, Vladimir Ilyich. **The Development of Capitalism in Russia: the process of forming the internal market for large-scale industry**. São Paulo: Nova Cultural, 1985 [1899] p.36-121.

LUCCI, Elian Alabi; BRANCO, Anselmo Lazaro; MENDONÇA, Cláudio. **Territory and Society in the Globalised World**. Vol. 02, 1 ed. São Paulo: Saraiva, 2010.

MANGOLIN, Cesar. The military dictatorship in Brazil: process, meaning and unfolding. **World Relations, Neoliberalism and Human Sociability.** 2 ed. São Bernado do Campo: Editora do Autor, 2011, p. 31-46. Available at:< http- cesarmangolin.files.wordprees.com-2010-02-cesar-mangolin-de-barros-a- ditadura-militar-no-brasil-2011.pdf> Accessed on: 20 January 2014.

MARQUES, Marta Inês Medeiros. **The Current Use of the Peasant Concept**. NERA Magazine. Presidente Prudente. Year 11. N° 12. p.56-67 . Jan.-June 2008. Available at: <http://revista.fct.unesp.br/index.php/nera/article/viewFile/1399/1381> . Accessed on: 13 June 2013.

MARQUES, Marta Inez M. **De sem-terra a 'posseiro', a luta pela terra e a construção do território peponês no espaço da Reforma Agrária: o caso dos assentados nas Fazendas Retiro e Velho - GO.** São Paulo: USP, 2000.

MARTINS, José de Souza. **The Captivity of the Earth**. 4 ed., São Paulo: Hucitec, 1990.

MARTINS, José de Souza. **Agrarian reform - the impossible dialogue on possible history. Tempo Social**; Rev. Sociol. USP, S. Paulo, 11(2): 97-128, Oct. 1999 (edited Feb. 2000).

MARX, Karl. **Capital**. vol.1. 8. ed. São Paulo: Difel, 1982.

MINAYO, M. C. DE S.; DESLANDES, S. F.; NETO, O. C.; GOMES, R. **Pesquisa social: teoria, método e criatividade.** Petrópolis, RJ: Vozes, 1994.

MINAYO, Maria Cecília de Souza. **Pesquisa Social - Teoria, método e criatividade.** 22 ed., Petrópolis: Vozes, 2003.

MINAYO, Maria Cecília de Souza. The challenge of social research. In: MINAYO, Maria Cecília de Souza; GOMES, Suely Ferreira Deslandes Romeu (eds.). **Pesquisa social: teoria, método e criatividade.** 27ª ed. Petrópolis: Vozes, 2008.

MINISTRY OF AGRICULTURE, LIVESTOCK AND SUPPLY. **Forestry Code.** 2011. Available at: <http://www.agricultura.gov.br/arq_editor/file/camaras_setoriais/Hortalicas/26RO/c artilhaCF.pdf> Accessed on 03 January 2014.

MINISTRY OF FOREIGN AFFAIRS. **Bolivia-Brazil gas pipeline - General Information.** 1999. Available at: <http://www.itamaraty.gov.br/sala-de-imprensa/notas-a- imprensa/1999/02/03/gasoduto-bolivia-brasil-informacoes-gerais> Accessed on: 14 March 2014.

MINISTRY OF AGRARIAN DEVELOPMENT. **Insertion of Agriculture Family Food in School Feeding.** Available at: <http://portal.mda.gov.br/portal/saf/arquivos/view/alimenta-o-escolar/MOC.pdf> Accessed 15 May 2013.

MOREIRA, Marcelo Carlos. **Agricultural Credit as an Element of the Agrarian Question: Analysing Credit in the State of Mato Grosso.** Dissertation (Master's) - Postgraduate/Master's Programme in Geography, ICHS, Federal University of Mato Grosso, Cuiabá, 2012.

MOREIRA, Marcelo Carlos. **Family Farming and Public Policies: The Implementation of PRONAF in the Corixinha Settlement - Cáceres/MT.** Cuiabá: UFMT, 2008.

MOREIRA, Ruy. **Where is geographical thinking going? Towards a critical epistemology.** São Paulo: Contexto, 2008.

MORENO, Gislaine. **Land and power in Mato Grosso: politics and mechanisms of cheating (1892-1992).** Cuiabá, MT: Entrelinhas/EdUFMT, 2007.

LANDLESS LABOUR MOVEMENT. **MST: Struggles and Conquistas.** 2010. 2nd ed. Available at: <http://www.mst.org.br/sites/default/files/MST%20Lutas%20e%20Conquistas%20 PDF.pdf> Accessed on 13 June 2013.

OLIVEIRA, Ariovaldo Umbelino de. **Peasant agriculture in Brazil.** São Paulo: Contexto, 1991.

OLIVEIRA, Ariovaldo Umbelino de. **Capitalist Mode of Production and Agriculture.** São Paulo: Atica, 1986.

PAULINO, E. T.. Territorial policies and the agrarian question: from theory to intervention. In: SAQUET, Marcos Aurélio; SANTOS, Roseli Alves. **(Agrarian Geography: territory and development.** 1.ed.São Paulo: Expressão Popular, 2010, v. , p. 107-129.

PAULINO, Eliane Tomiasi; ALMEIDA, Rosimeire Aparecida de. **Terra e Territótios: a questão camponesa no capitalismo.** São Paulo: Expressão popular, 2010.

PCBAP - Upper Paraguay Basin Conservation Plan. **Pantanal Project, National Environment Programme.** 1997. v. II, tome III (Diagnosis of the physical and biotic environments). Brasilia: PNMA.

PINTO, R. S.; FIGUEIREDO, D. S.. **Formation of the Bolivian agrarian space.** NEAF Texts, Postgraduate Programme, p. 1 - 33, 30 May 2011. Available at: http://mafds.websimples.info/files/arquivo/92/Texto_22.PDF Accessed on: 12 June 2013.

POTT, A. . The **relationship between vegetation and environments in the Pantanal.** In: Meeting on Remote Sensing Applied to Studies in the Pantanal, 1995, Corumbá, MS. Book of Abstracts, Meeting on Remote Sensing applied to studies in the Pantanal. São José dos Campos, SP: INPE, 1995. p. 5-7.

RESENDE, Raélita de Oliveira. **Study of the *Curatella Americana* l. Species (Lixeira) Used as a Bioindicator in the Auriferous Region of the Cangas District in Poconé / MT.** Postgraduate dissertation in Geosciences, Institute of Exact and Earth Sciences, Federal University of Mato Grosso, 2012.

JOURNAL OF THE HISTORICAL AND GEOGRAPHICAL INSTITUTE OF MATO GROSSO. Onélia Carmem Rossetto and Antonio C. P. Brasil Jr. **Entre Cheias e Vazantes: Historical Characteristics of the Occupation of the Mato Grosso Pantanal.** Vol 59, 2001.

RIBEIRO, E. M. ; GALIZONI, F. M. ; MELO, A. P. G. ; MOREIRA, T. M. B. ; ALMEIDA, A. F. C. S. ; CARVALHO, A. A. ; CALDAS, A. L. T. . **Local Markets, Rural Domestic Industry and Commercialisation in Family Farming in Alto Jequitinhonha**. In: XV Seminar on the Economy of Minas Gerais, 2012, Diamantina. Proceedings of the XV Seminar on Mining Economics. Belo Horizonte: Cedeplar/ UFMG, 2012.

RIBEIRO, Luiz Cesar de Queiroz; SANTOS JUNIOR, Orlando Alves do (eds). **The Metropolises and Brazilian Social Issues.** Rio de Janeiro: Revan, 2007.

RIBEIRO. Vanderlei Vazelesk. **La Roçay la Campana: the agrarian question under Varguism and Peronism in a comparative perspective.** Thesis (doctorate), Fluminense Federal University, Niterói, 2006.

RICHARDSON, Roberto Jarry. **Social Research: methods and techniques.** 3. ed. São Paulo: Atlas S.A, 1999.

ROCHA, Ronaldo dos Santos da; Celestino, Vivian da Silva. **History of the territorial occupation of Brazil.** Recife: UFRS, 2010.

ROSSETTO, O. C. ; BRASIL JÚNIOR, A. C. P. . **Between Floods and Troughs: Historical Characteristics of the Occupation and Sustainability of the Pantanal of Mato Grosso.** Revista do Instituto Histórico e Geográfico de Mato Grosso, Cuiabá - MT, v. 59, p. 91 -112, 2001

ROSSETTO, Onélia Carmem; GIRARD, Eduardo Pulon. Agrarian dynamics and socio-environmental sustainability in the Brazilian Pantanal. **Revista Nera**, Presidente Prudente, São Paulo, n° 21, p. 135-161, Jul.-Dec. 2012.

SHANIN, Teodor. **The Definition of Peasant: Conceptualisations and Disconceptualisations - the Old and the New in a Marxist Discussion**. Revista NERA, Presidente Prudente, Ano 8, n. 7 p. 1-21 Jul./Dec. 2005.
Available at:
<http://revista.fct.unesp.br/index.php/nera/article/viewFile/1456/1432> Accessed on: 4 May 2013.

SCHMIDT, Mario. **New Critical History.** Ed. Nova Geração.1996.

SCHWENK, L. M. Biogeographic Domains. In: MORENO, G.; HIGA, T. C. S. **(Geography of Mato Grosso: territory, society, environment.** Cuiabá: Entrelinhas, 2005. p. 250-271.

SILVA, Ana Cristina Vieira. **Settlement 14 de Agosto Campo Verde - MT: economic and social characteristics.** Cuiabá: UFMT, 2009.

SILVA, C. A. **Análise da Estrutura Fundiária Mato-grossense.** Cuiabá: UFMT, 2011.

SILVA, José Roberto Pereira. **Study of Occupation and Socio-environmental Conflicts in the Settlement and PPAs of the Igarapé do Bruno - Apiacás - MT.** Cuiabá: UFMT, 2008.

SILVA, L. R. C. ; DAMASCENO, A. D. ; MARTINS, M. C. R. ; SOBRAL, K. M. ; FARIAS, I.M.S. . Documentary research: an investigative alternative in teacher training. In: IX Congresso Nacional de Educação - EDUCERE/ III Encontro Sul Brasileiro de Psicopedagia, 2009, Curitiba. **Proceedings of** the IX National Congress of Education - EDUCERE [electronic resource]: Educational Policies and Practices:

challenges of learning; **Proceedings of** the III South Brazilian Meeting of Psychopedagogy. Curitiba: Champagnat, 2009. v. 1. p. 4554-4566. Available at: <http://www.pucpr.br/eventos/educere/educere2009/anais/pdf/3124_1712.pdf> Accessed on: 06 October 2013.

SILVA, Thiago Moreira Melo e. **A Presença das Ligas Camponesas no Nordeste.** São Paulo: PUC/SP, 2009. Available at: <http://www.geografia.fflch.usp.br/inferior/laboratorios/agraria/Anais%20XIXENGA/artigos/Silva_TMM.pdf> Accessed on : 08 August 2013

SIMONETTI, M. C. L.. **The Long Walk: (re)construction of peasant territory: the struggle for land and settlements in Brazil.** PhD Thesis - Department of Geography, University of São Paulo, São Paulo, 1999.

STEDILE, J. P. (org.) **A Questão Agrária no Brasil: Programas de reforma Agrária 1946 - 2003.** São Paulo: Expressão popular, 2005.

TRIVINOS, Augusto N. S., **Introdução à pesquisa em Ciências Sociais.** São Paulo: Atlas, 1992.

VIEIRA, Luiz. M. **Pantanal: a threatened biome.** 2002. Available at: <http://www.agronline.com.br/artigos/pantanal-um-bioma-ameacado> Accessed on 25 April 2013.

VILA DA SILVA, J. S; ABDON, M. M. **Delimitation of the Brazilian Pantanal and its Subregions.** Revista Pesquisa Agropecuária Brasileira. Brasília, v. 33. n°. especial, p.1703-11, oct. 1998.

Viladesau, Tomás Palau. **The soya agribusiness in Paraguay Antecedents and social and economic impacts.** In: Landim, Regina Ângela Bruno - Campesinato e Agronegócio na América Latina. São Paulo, Expressão Popular, 2008, p. 17-43.

VILARINHO NETO, Cornélio Silvano. **Regional Metropolisation, Formation and Consolidation of the Urban Network of the State of Mato Grosso.** Cuiabá: EdUFMT, 2009.

WAHREN, J. Construyendo Territorios Corporaciones, Movimiento Social y Proyectos Autogestionados en Mosconi, Salta (Argentina). In: FERNANDES, B.

Peasantry and Agribusiness in Latin America: the Current Agrarian Question. São Paulo: Expressão Popular. 2008. p. 385-401.

<http://www.mda.gov.br/portal/> Accessed 26 April 2013.

< http://www.incra.gov.br/index.php/institucionall/historico-do-incra> Accessed 26 Apr. 2013.

<http://www.mteseusmunicipios.com.br/NG/indexint.php?sid=199> Accessed on: 03 October 2013.

ANNEXES

I want morebooks!

Buy your books fast and straightforward online - at one of world's fastest growing online book stores! Environmentally sound due to Print-on-Demand technologies.

Buy your books online at
www.morebooks.shop

Kaufen Sie Ihre Bücher schnell und unkompliziert online – auf einer der am schnellsten wachsenden Buchhandelsplattformen weltweit! Dank Print-On-Demand umwelt- und ressourcenschonend produzi ert.

Bücher schneller online kaufen
www.morebooks.shop

Printed by Books on Demand GmbH, Norderstedt / Germany